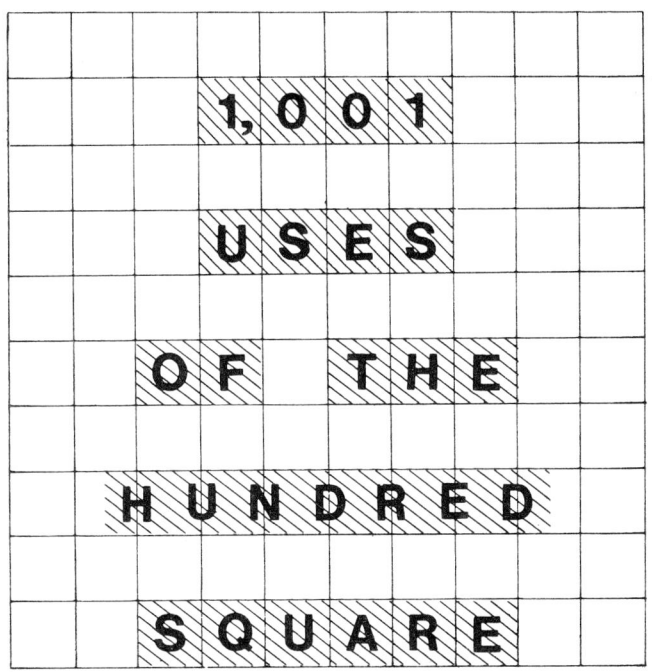

1,001 USES OF THE HUNDRED SQUARE

ACTIVITIES and IDEAS
for Teaching Mathematics

Leah Mildred Beardsley

ILLUSTRATIONS by the AUTHOR

Parker Publishing Company, Inc. West Nyack, New York

This book is dedicated
with love
to
Andy,
Anne Leah, Joan Ellen, and *John Bruce*

©1973 by

Parker Publishing Company, Inc.

West Nyack, N.Y.

All rights reserved. No part of this book may be reproduced in any form or by any means, without permission in writing from the publisher.

Library of Congress Cataloging in Publication Data

Beardsley, Leah Mildred,
   1,001 uses of the hundred square.

   1. Mathematics--Study and teaching--Audio-visual aids. 2. Mathematics--Charts, diagrams, etc. 3. Square. I. Title. II. Title: The hundred square.
QA19.C45B4   372.7'3   73-6786
ISBN 0-13-636910-3

Printed in the United States of America

How This Book Will Help You Teach Math More Effectively

"1,001," a legendary number, was chosen for the title because a child explained that it means "lots and lots and lots," and that is a good estimation of the number of ideas you'll find in this book.

The hundred square is an empty 10 by 10 grid, but it's also a multi-purpose aid that will fulfill a great many of your teaching needs.

As a mathematics demonstration teacher in Detroit, I visit many Elementary and Junior High schools and advise and help hundreds of teachers and thousands of boys and girls. My main job is demonstrating math lessons on topics requested by teachers, in their own classrooms. The 100-square provides the visual aids I use most frequently to introduce a lesson, illustrate or validate an example, motivate practice for speed and accuracy, explain a property, stimulate and sustain interest, change a mood, provide a challenge, wake up a sleepy student, or otherwise involve the class in an effective learning situation.

Third-grade children love the Commutative Law of Multiplication Chart, (page 123 — where the King's Men march in an array, then turn and go up a hill without losing a man. Sixth graders understand one of the reasons for using the Distributive Property of Multiplication (page 127) when they see an array that does not fit on the Multiplication Table Chart, and know that to work the problem they must either use the distributive law or learn the 14th table, (or the 26th, the 47th, or the 196th, and never come to the end of learning tables — such a wonderful, important law for them to have).

This book contains many other activities for all grade levels from Kindergarten through Junior High. It is intended to be used as a resource book to supplement regular textbook work. The design theme was selected for the first chapter because many teachers feel that learning numbers should not be the child's first experience in the learning of mathematics. They realize that the action of classifying objects, perhaps by color, or shape, or size, or the experience of continuing a pattern in design that involves these attributes, requires real understanding on his part. This type of logical thinking will be a great asset for him in all his mathematical work, and many challenging activities for developing logical awareness are included early in the book.

Story problems are an important part of every child's math education. These 100-square visuals may be incorporated in your story problem work. Conversely, number stories may be created for nearly all the visuals in the book. Examine the inchworm problems in the measurement chapter, the cotton problems in the graph section, the rooster problems in the addition chapter, and paving the road in the fractions chapter. To individualize your story problems, use your children's names in the stories, let their interests and the seasons of the year help make the problems especially theirs, capitalize on the real problems in the classroom – amount of paper needed, lack of pencils (always a problem, isn't it so?), books matched to children, telephone numbers, lunch money, and the average heights of your boys and girls. Again, let your 100-square visuals help solve these problems. (And if you ever solve the pencil problem, take a bow.) Encourage children to employ the technique of thinking through each problem situation, and of estimating the answer before beginning the work.

The ideas suggested in the book have been tried out in many classrooms. Delighted teachers tell me, that when used with a definite purpose, they have helped to develop, clarify, and extend student understanding in many basic areas. They report that the mathematical games are captivating, even if Wolf never captures Little Pig. They suggested a book so that mathematics teachers everywhere could benefit, and so I share my ideas with you, and believe you will like them and find them practical.

Simply stated: "The simple things in life are best." Experienced teachers realize this, and seek *simple* but effective ways to illustrate the abstractions of mathematics.

The 100-square is a *simple* thing. What are its reasons for being a best thing?

> 100-square visuals are *quickly and easily constructed*
>
> and help to *bridge the concrete-abstract span*

in *innumerable* ways

by *assisting* the questing teacher

to make mathematics more *meaningful for students*

at *every grade level*

in every basic math *area*

making any lesson an *enrichment* experience

saving valuable hours of the *teacher's time*

at little or *no expense*

This book

provides the workable *ideas*

shows a *picture* of each visual aid

suggests *uses* and guidelines

gives samples that *explain how and why* the aid works

The ideas offered are designed for

teacher-class large-size charts

individual pupil grids

bulletin board displays

overhead projector tranparencies

individual manipulative activities

games that teach, review and drill

May I recommend the 100-square as a remarkable learning device and a most practical aid for *all* mathematics teachers.

Leah M. Beardsley

Acknowledgements

Many thanks to all the teachers who gave me the chance to try out these ideas with their children, and who added functional, helpful suggestions.

Many thanks to all my friends for their encouragement and confidence, and especially to Miss Irene Sauble, Mrs. Muriel Greig, Dr. Geraldine Green, Miss Mary Durgan, Dr. Theresa Denman, and Miss Erma Kellogg.

Many thanks to Dr. Eugene P. Smith, President, 1972 National Council Teachers of Mathematics, for suggesting the title.

My deep appreciation to my daughter, Joan, who spent countless hours helping to type the manuscript.

Contents

How This Book Will Help You Teach Math More Effectively • 3

1 MATH ACTIVITIES – A STUDY IN DESIGN • 15

—imaginative enrichment experiences lead to practical mathematical awareness

Fun With Color Patterns—leads to number sequences •	18
Missing Links—a readiness activity—sequence •	19
Block Logic Fill-Ins—finding clues to solve problems •	20
Shapes and Colors—classification •	20
No Row Nor Column Alike—differences •	21
Parquetry Arrangements—develop discrimination •	22
Symmetry In Square Arrays—balancing •	23
Frieze In Block Pattern—resembles a graph •	24
Border Designs: Initials, Math Symbols, Cross Stitch •	24
Diamonds and Sparkle—repeating a pattern •	25
Modern Effects—geometric abstraction—percents •	25
Rod Arrangements—design your own—size and color •	26
Geometric Block Designs Help Teach Number Theory •	27
Odd Number Sequence	27
Square Number Progressions	27
Line Segments on Squared Paper—basic geometry •	28
Curved Path From Line Segments forming parabolas •	
Circle Symmetry—skill in compass use •	29
Create A Follow-The-Number Design—counting order •	30

2 THE VISUAL IMPACT OF GRAPHS • 33

—information, organization, representation

A Birthday Graph That Each Child Helps To Construct—bar, tally, or pictograph •	36
Graphs For Meaning of Number •	37
Graphs For Practicing Facts •	38
Graphs For Keeping Records •	38
Class Achievement Record •	39
Individual Pupil's Item-Analysis Graph •	40
Graphs For Introducing Lessons •	41
Combination Temperature—Rainfall Graphs •	41

2 THE VISUAL IMPACT OF GRAPHS *continued*

Graphs For Comparison •	43
Experimenting With Line Graphs—Broken Line Graph •	44
Experimenting With Line Graphs—Batter-Graph •	44
Suitable and Purposeful Graphs For Various Grade Levels •	46

3 UNDERSTANDING NUMBER SYMBOLS • 49

—symbols have a high level of abstraction—first build a high level of understanding

Equivalent And Nonequivalent Sets—one-to-one matching of pictures, colors, dots •	51
Counting And Number Order—squared activities are designed for practicing ordinal concepts •	52
Number Symbols And Pictures •	53
Number Symbols And Dots •	54
Number Symbols, Number Words And Squares •	54
Practice Your Figure 8's—individual practice cards •	56
Ever-Ready Bulletin Board Number Practice Chart •	56
Place Value—Building A Ten •	57
Count 1,000 Squares By 100's •	58
Arrow Moves—A Pre-Number Line Bulletin Board Chart •	58
Your Fingers Do The Walking •	59
Counting Rows And Number Lines •	60
March Left, Right, Up And Down—children give orders •	60
Number Lines From Squared Paper •	61
Absolute Value On An Integer Number Line •	62
Slide Rule Number Lines—whole numbers, fractions, integers and other base slides •	62

4 BRIGHT IDEAS FOR USING THE COUNTING CHART • 65

—a welcome adjunct to every math room

The Counting Chart—a compact picture of numbers from 1 to 100 •	68
Variation Of The Counting Chart •	68
Removable Numerals •	68
A Counting Chart From 100 To 200 •	69
Using The Counting Chart—recognition of numbers, importance of ten, counting by 10's, number order, missing numerals, before, between, after, vocabulary, greater than, less than, pattern •	69
Adding One and Adding Ones—commutative law, the difficult teens, cut and paste, subtracting ones, adding 10, adding 9, adding 20 and 19, subtracting 10, subtracting 9, finding addition and subtraction patterns and interrelationships, families help with the facts •	73
Magic Sums—and the Inverse Magic of Subtraction •	76

4 BRIGHT IDEAS FOR USING THE COUNTING CHART *continued*

Multiplication And Division Exercises On The Counting Chart—multiple patterns 2nd table, even and odd numbers, 5th table, even division, uneven division, 9th table, 3rd table, 6th table and primes, 15th table, vertical multiple patterns, 4th table •	78
Square Numbers •	84
Eratosthenes' Sieve For Finding Primes •	85
Counting Chart Patterns •	86

5 SOLVE YOUR ADDITION AND SUBTRACTION PROBLEMS • 89

—stress "place value", there are pitfalls in regrouping

Dominoes •	92
Open-Up Flashcards •	92
Subsets Teach Families of Facts •	93
Subtracting From 5 •	93
Picture Arrays •	94
Inverse Operations •	94
Show-Me-Strips Help Solve Number Stories •	95
Addition-Subtraction Chart •	96
Constructing Individual Addition Charts—use an L-slide, correlate addition and subtraction •	96
Addition and Subtraction Practice Strips •	97
Adding to Specific Sums •	98
Pictures Check the Facts •	99
Construct A Tens Chart—ten is our important number •	100
Stars And Zigzags—addition facts •	100
Long Zigs And Short Zags—writing subtraction problems •	101
Triangles and a Pair Of Those Crooks—three addends •	101
Are You a Commutative Adder—practice with rows of bright colored squares •	101
Associative Property for Addition—color sets of squares •	102
Calendar—its a kind of counting chart—use it for addition practice •	103
Ten Is The Key—Place Value and Regrouping •	104
Flashcards That Children Construct For Their Own Needs •	105
Place Value and Addition •	105
Tens And Ones—relates squares to subtraction with regrouping •	106
Counting Row Addition •	107
Counting Chart Addition •	107

6 HAPPY MULTIPLICATION TO YOU • 109

—having a multiplication-table struggle? Is long division often "wrong division"? Here's lots of help

Introducing Multiplication—Five Approaches: equivalent sets, repeated addition; array; number line; cross product •	112

6 HAPPY MULTIPLICATION *continued*

A Bulletin Board Project—Multiplication by repeated addition of equivalent sets; Noah's Ark, Baby Elephant's Trunk, Little Caterpillar, Short Train, 6-Inch Ruler, Little House, Short Chain, Candles, and Colorful Little Rooster •	115
Multiplication-Division Chart—the Magna Charta of all charts; construct by counting with slides or arrays •	118
Using The Multiplication Chart •	120
Introducing Division—without or with remainders •	120
Finding Averages—a new approach •	121
Story Problem—learn to pretend •	121
Exploring Division—Never Divide By Zero •	122
More Helps For Conquering the Multiplication Facts—build a staircase; see the King's Men march to show off the Commutative Property; meet the Identity Element for multiplication; learn the pattern of the finger table, the 9's; set apart the difficult facts as you cover the table with the help of the laws •	123
Discover Why The Distributive Property Is The Great Law Of Multiplication— a simple array makes it clear •	126
Introducing the Multiplication Algorithm—children appreciate shortcuts •	127
Fingers Walk On The M-D Chart—show the distributive law •	128
Estimating When Solving Story Problems •	128
Introducing 2- and 3-Place Multipliers •	129
Distributive Property Practice •	130
Think Distributive—when you do l-o-o-o-o-ng problems •	131
Squaring A Binomial •	131
Perceive the Importance of Place Value When Multiplying and Dividing •	133
Long Division, Not Wrong Division—divide sets of squares •	133
Windows Give a Clear View of the Associative Property—a variety of bulletin board ideas meaningfully explains the Associative Law •	134
Square Numbers, and One-Off-The-Square •	136
Factoring Activities •	137
Determining Primes and Composites •	137
Lattice Multiplication •	138
Napier's Bones •	139

7 MEASUREMENT • 141

—a search for the exact

Does The Hat Fit?—a scale for measuring your pupils heights •	144
Feet In A Mile—little feet give the clue •	146
Cups To Gallons—Comparison of Units •	146
Area Is Square Measurement—Perimeter Is Line Measurement •	146
The Climbing Inchworm—a fun problem •	147
Square Centimeter, Square Inch •	147
Longer Trips For Inchworm •	148

Contents

7 MEASUREMENT *continued*

Four Square Units—Bigger Squares Out Of Small Squares •	149
Standard Area Measurement—square feet, square yard •	149
Building A Pen For The New Pet—comparing perimeter and area—you'll wonder where the area went •	150
Big Boys—which is bigger Hi Boy or Lo Boy? •	152
Area Cards Teach Algebra •	153
Sweeping The Sidewalk Reveals Its Area •	153
Triangular Area—Formula Discovery—sweep the triangles •	154
Areas That Do Or Don't Fold Around A Cube—test your perceptive abilities •	156
Lost Area Puzzle •	156
Same Dog—Different Scale •	157
A Circle In A Square—hints on finding the area and the formula •	159
A Giant Protractor •	159
Surface Area Of A Prism Means Finding Area Of 2 Bases And 4 Lateral Faces—a laboratory approach •	160
Tessellations On Squared Paper—some shapes fit around a point and some gap or overlap •	161
Octagons With Square Joiners •	162
Polyominoes—you know dominoes, meet the others •	163
Tessalate Pentominoes and Trominoes •	163
Pythagorean Theorem—a square problem •	164
The Square Root of 2—use squares to locate it on the number line •	165
Fahrenheit To Centigrade—or vice versa—how to do it and remember how •	165

8 FRACTIONS—BITS OF THIS AND THAT • 167

—action is a big part of <u>fractions</u>

Color-Squared Problems To Make You Think—numerators and denominators •	170
Identifying Fraction—action work with hundred squares, scissors and crayons •	171
Construct a Fractional Addition-Subtraction Chart •	171
Fractions Of Sets •	172
Do-It-Yourself •	172
Fractional Equivalence Chart—use it for adding unlike fractions •	173
Fractions On The Multiplication Chart—it's full of fractions—and explains raising to higher terms •	173
Use The Multiplication Chart To Find Common Denominators •	174
Crisscross—equivalent fractions •	175
Coordinate Graphs of Equivalent Fractions—where are the proper fractions? •	175
A Line You Must Cross—fractional equations •	176
Colored-Squared Subtraction Problems With Regrouping—when a fraction needs a friend •	176

8 FRACTIONS—BITS OF THIS AND THAT *continued*

United States Currency—money to work with •	177
Fill A Pig—different ways to bank your money •	178
Decimal Place Value—Whole Numbers and Fractions •	179
Percents And The Hundred Square—graphic models of percentage •	179
Pave A Road—use story problems and number lines •	182
Pave A Road—use number line for all fractional problems •	183
Fractional Area—colored-squared multiplication •	183
Construct A Multiplication-Division Chart For Fractions •	184
A Fraction Of A Fraction—cut the squares •	184
Reciprocals—can you picture them? •	194
Distributive Law Arrays—apples, colors and plain squares illustrate the Distributive Law holds for multiplication of fractions •	185
Comparison Puzzles •	186
Great Discovery—find out why Euclid's Algorithm for greatest common divisor works so well •	187
Now Have Fun—Be Different As You Rename One-Half—Crazy Fraction Multiplication Chart •	188
A Whale On A Scale, and A Snail On A Scale •	189

9 IT'S YOUR TURN—GAMES FOR WORK AND PLAY • 191

—the boys usually beat the girls—playing baseball

Poison Square •	193
Wolf And Little Pig •	194
Go •	195
The Bear Went Over The Mountain •	195
Lucky Tens •	196
Rule Pictures •	196
Around The Block •	197
A Long Space Trip •	197
Jumper, The Cricket •	197
Zip-O's—Add-0, Mult-0, Fact-0, Add-0-Win, Productivity, Fraction-Action •	197
Football •	199
Baseball •	199
Roll-Overs •	199
Square Nims •	200
Battleship •	201
Park The Cars •	201
Math Words •	202
Sign Game •	202
Follow Me Game •	203
The Mouse Ran Up The Clock •	203
Hopscotch Multiplication •	203
Ferris Wheel •	204
Race Track •	204
Frame Game •	205
Magic Squares •	205

Contents 13

9 IT'S YOUR TURN—GAMES FOR WORK AND PLAY *continued*

Tic-Tac-Toe •	206
Pebble Toss •	206
Plant The Corn •	208
Climb Up And Down The Ladder •	208
Climb Up, Slide Down •	208
Add–Subtract–Spell •	209
Whiz-Bang Column Addition •	209
A Square Man •	210
Trucks Loaded With Rectangles •	211
Make An Equation—Up, Down, Across, Diagonal •	211
Equation Mix-Up •	212
Can You Build Numbers? •	212
Grand Total—Discover and Explain •	212
Extensions of Grand Total •	213
Checker Move: Least Common Multiple or Denominator •	214
Cross Puzzles •	215
Secret Words •	215
Fraction Code •	215

10 ODDS AND ENDS • 217

—locate a school, graph an equation, let a computer think for you, make a point

Other Bases	
Change-A-Base Counting Chart •	220
Base 5 Counting, Addition, and Multiplication Charts •	221
Base 5 Place Value •	222
Convert Base 10 to Base 5 •	222
Secret Number Cards •	223
Electronic Computer •	225
Multiplying Powers of 2—add exponents •	225
Coordinates •	226
A Trip To School •	226
Tic-Tac-Toe On The Coordinate Grid •	226
Map Coordinates •	227
Symmetry And Reflections On A Coordinate Grid •	228
The Grab-Bag Game And Solution Sets •	230
Graphing the Grab-Bag Equation—a linear equation •	230
Where Do You Sit?—Inequalities •	231
Graphing Simultaneous Equations •	231
Intergers Multiplication Patterns •	232
Probability •	233
Penny Flip •	233
Math Luck •	233
The H and T Game •	235

Index • 237

1.

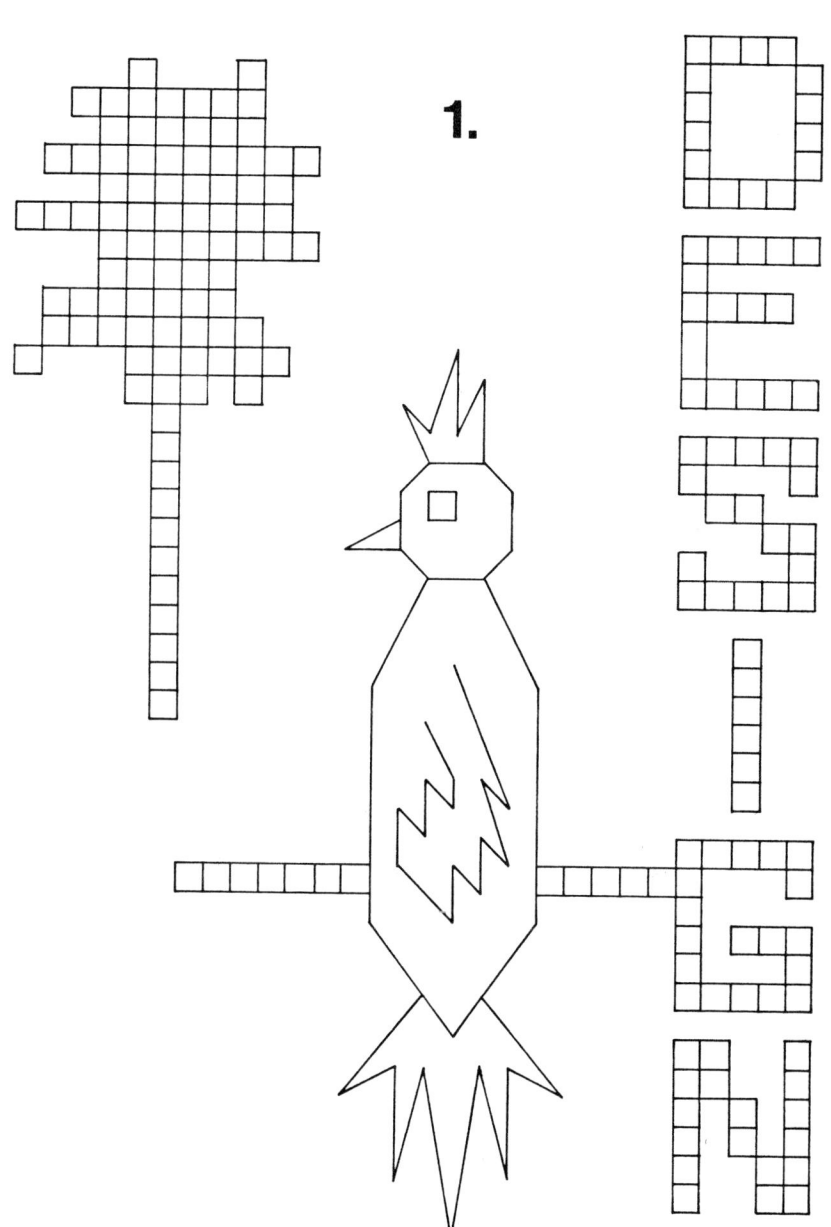

1

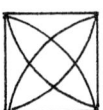 # Math Activities
A Study in Design

—imaginative enrichment experiences lead to practical mathematical awareness

This book begins with a study of design, for this creative and perceptive experience results in heightened mathematical skill. The ability to perceive and conceive designs requires logical thinking and discrimination of judgment, and makes necessary the exploration of basic shapes, patterns and symmetry.

Graphs, charts, geometric drawings, and other structured forms are often only certain extensions of design. Not so readily perceived is the sense of order and pattern that is interwoven throughout the entire field of mathematics. But it is the discovery and constant rediscovery of this unifying design that fascinates the student and fills him with wonder.

The following discussion of the terms design, pattern, and properties as they are used in natural science, helps capture the interest of the students and prepares them for the use of such terms in their mathematics and science studies.

For millions of years, nature has employed, tested, sifted out, and rejected infinite numbers of nonfunctional designs trying to produce an ordered universe. For each retained element and life form there exist natural properties. When man discovers and understands these he can make predictions and put them to use.

There are many evident examples of these natural laws. Oxygen supports life. Our moon men knew of this natural property and made their trip easier by taking oxygen with them to the almost airless moon. Hydrogen is lighter than air. Some years ago, large dirigibles were filled with hydrogen, rose in the sky and sailed in silver beauty over land and sea. But when another property of hydrogen, its high flammability, caused a dirigible to burst into flame, the use of this gas for inflating these buoyant airships was discarded in favor of helium, with its natural properties of lightness and nonflammability.

One of the basic shapes in nature is the triangle. Man has discovered the great inherent strength of this shape, and builds radio towers, strutted into many triangles, for the strength necessary to support their great height.

Bears have a natural property that the children in your classroom do not have — they can grow their own fur coats. If you see a bare bear it isn't a bear. Monkeys can swing by their tails — wonderful for tree-dwellers. Children can grow taller — adults can't. These few examples help to explain what is meant by a natural property.

Artists study mathematics, make use of the same shapes and dimensions as the mathematicians, and search for patterns and natural properties of design when creating up-to-date usable forms. Buildings, cars, planes and trains, spaceships and sea ships, furniture and children's toys are all shaped to be both functional and pleasing to the eye.

As the scientist, the artist, the builder and the engineer need to understand related designs in their fields, so the mathematics student should search for and understand the related patterns in the mathematical system he uses, and make use of the natural properties of operations and number when he solves his problems. You, the teacher, will wish to provide many opportunities for your students to investigate and discover these interrelations and apply them with understanding to math situations.

A study in design is not intended simply to teach children to draw well, or to keep them pleasantly busy. It has an immediate and also a long-lasting effect on their entire mathematics program.

Design creation may be simple or complex, traditional or modern, copied or original. The study may be begun and carried on with enthusiasm in the kindergarten, and continued throughout the secondary grades whenever and wherever students are encouraged to express themselves freely. The examples shown in this book are merely representations of the many possible forms that may be utilized to give purpose to the study of design. Each selection was chosen to emphasize a definite mathematical idea. Supplement these suggestions and include your own creative extensions to prepare delightful experiences for you and your students.

Figure 1.1

FUN WITH COLOR PATTERNS – TEACH SEQUENCE

Patterns permeate much of our mathematics. Arithmetic reveals number patterns repeating in sequence. In geometry, shapes form artistic patterns.

Kindergarten and primary children may copy or create simple color patterns in a row of squares. The three samples in Figure 1.2 have only two different colors or forms. Each of these will be said to have a "one-two" pattern. This pattern is pictured numerically in Figure 1.3

Figure 1.2

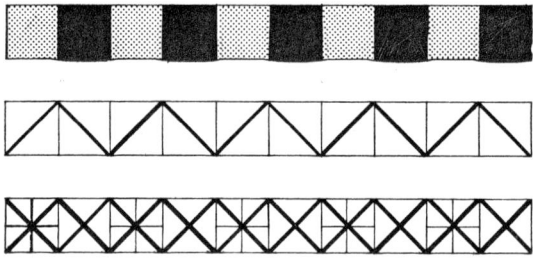

Figure 1.3

| 1 | 2 | 1 | 2 | 1 | 2 | 1 | 2 | 1 | 2 |

Math Activities—A Study in Design

A more difficult pattern makes use of three colors and needs three numbers to illustrate the changes. See Figure 1.4

Figure 1.4

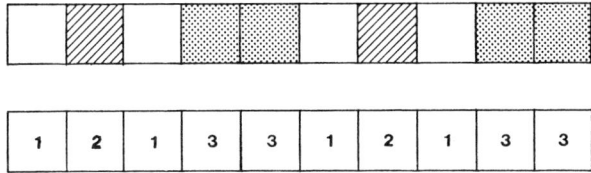

MISSING LINKS — COLOR AND NUMBER SEQUENCE

Have the children tell what form or color should go into each blank square to complete the pattern chain of each row in Figure 1.5.

Figure 1.5

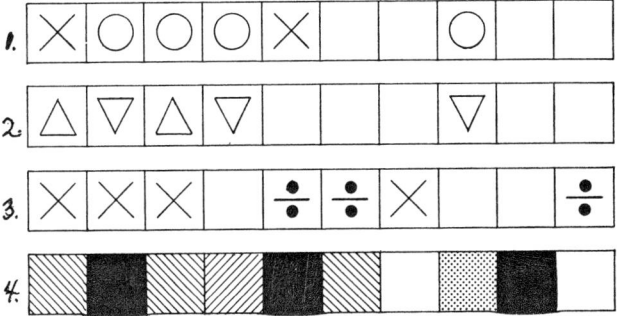

Using rows of blank squares, have the children write in the number patterns that correspond with each row in Figure 1.5. Figure 1.6 shows these corresponding number patterns completed. This exercise is a readiness activity for number sequence.

Figure 1.6

	1	2	2	2	1	2	2	2	1	2
1.										

	1	2	1	2	1	2	1	2	1	2
2.										

	1	1	1	2	2	2	1	1	1	2
3.										

	1	2	1	3	2	1	3	4	2	1
4.										

BLOCK LOGIC FILL-INS – LOGICAL THINKING

What comes next? Pupils are to discover the pattern and correctly fill in each block until the row is completed. An exercise of this type is to be used either as a test of logic ability, or to help sharpen awareness and discrimination. Finding the clue to each geometric pattern simulates finding clues in subsequent number sequences (Figure 1.7).

Figure 1.7

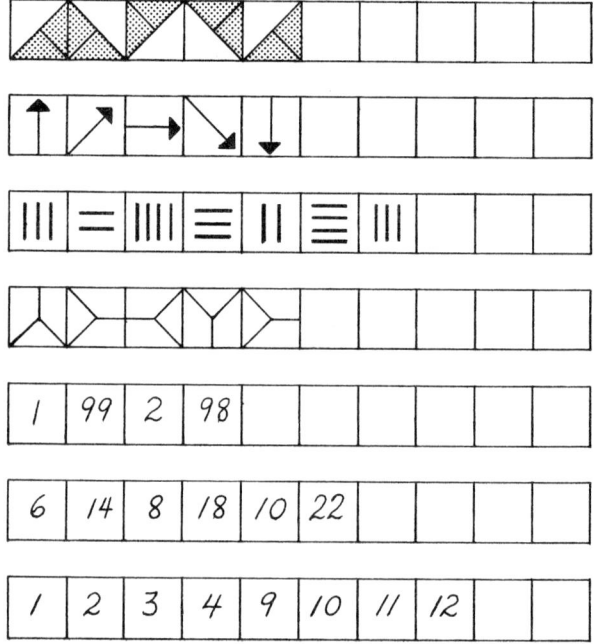

SHAPES AND COLORS – CLASSIFICATION

Use these nine blue, red, and yellow geometric forms to complete these problems One row of blank squares is needed for each problem (Figure 1.8).

Figure 1 8

1. Study the geometric forms that have already been drawn, and then continue the pattern until each of the nine forms has been used once (Figure 1.9).

Math Activities—A Study in Design

Figure 1.9

2. In the first block color the blue circle. In the second block color a geometric form different in *one* way from the blue circle. (This may be a blue square, a blue triangle, a red circle, or a yellow circle.) In the third block color a form different in *one* way from the second, and so on. Try to use as many of the nine forms as you can. Figure 1.10 shows two completed examples.

Figure 1.10

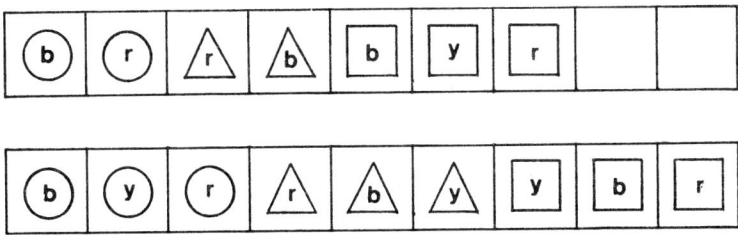

3. In the first block color the red triangle. In the second block color a form different in *two* ways from the red triangle. (A blue circle, a blue square, a yellow circle or a yellow square would be correct.) The third block must be different in *two* ways from the second block, and so on. Try to use all nine geometric forms. See Figure 1.11 for completed examples.

Figure 1.11

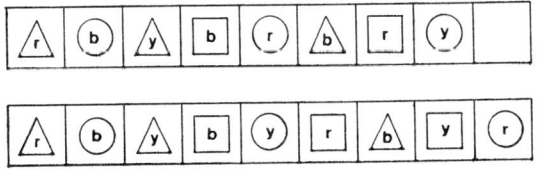

NO ROW – NO COLUMN ALIKE – DIFFERENCES

Arrange nine geometric forms (three shapes, three colors) in a 3 x 3 array so that in no row nor column will there be any two shapes or colors alike. Figure 1.12 shows a possible arrangement.

Figure 1.12

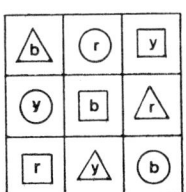

Using construction paper cut four different colors of each shape shown in Figure 1.13, making 16 pieces altogether. Place these shapes in a 4 x 4 array of squared paper, so no two colors or shapes are alike in any row or column. Figure 1.14 shows a completed example.

Figure 1.13

Figure 1.14

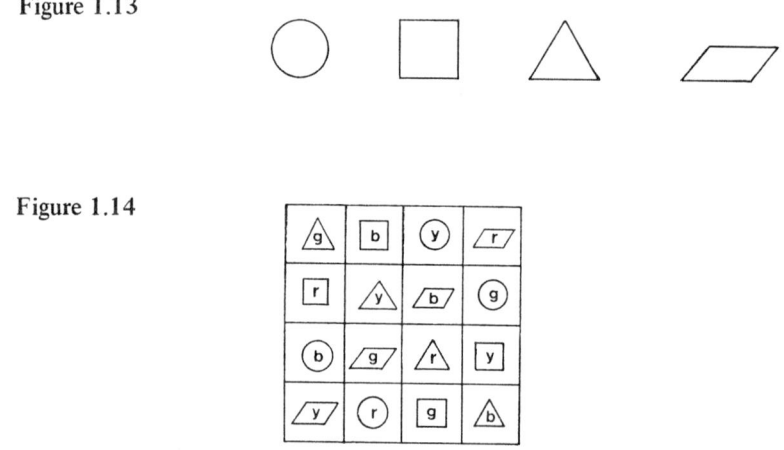

Bulletin Board Idea: Using a 5 x 5 array of large squares, a committee will work out an arrangement of five different shapes and colors (Figure 1.14).

PARQUETRY ARRANGEMENTS – DEVELOP DISCRIMINATION

Prepare a set of squares similar to the set shown in Figure 1.15, and sample design cards of possible arrangements, as in Figures 1.16, 1.17, and 1.18. These cards should have no guide lines, and for further difficulty may be drawn in a different scale from the squares. The objective is to form a copy of the design shown on the example card. Also, challenge the children to create their own interesting arrangements and draw them on cards for their classmates to copy.

Figure 1.15

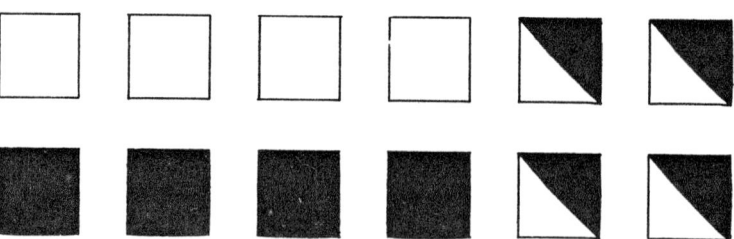

Figure 1.16

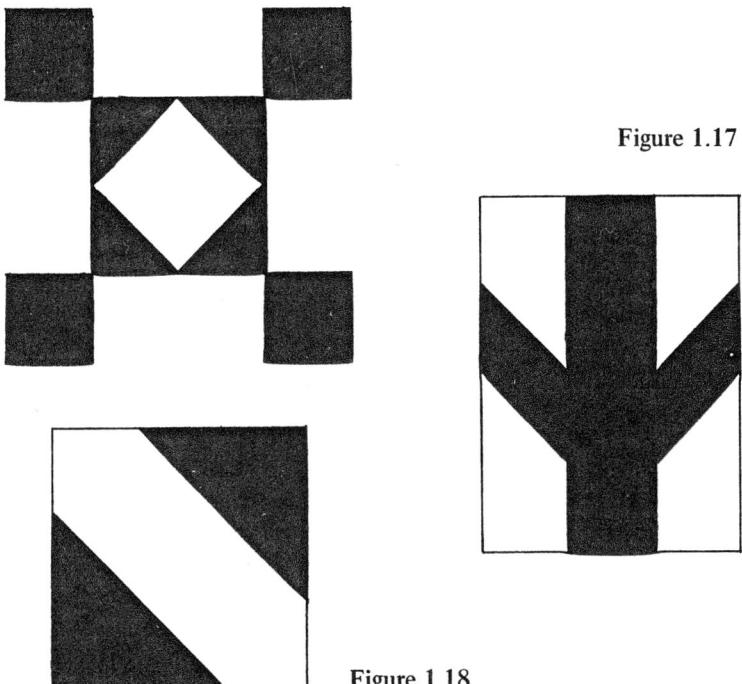

Figure 1.17

Figure 1.18

SYMMETRY IN SQUARE ARRAYS – BALANCING RELATION

Decorative compositions are colored in various size grids. Study the examples and then design your own. Be original (Figures 1.19, 1.20).

Figure 1.19

Figure 1.20

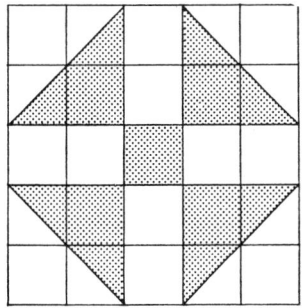

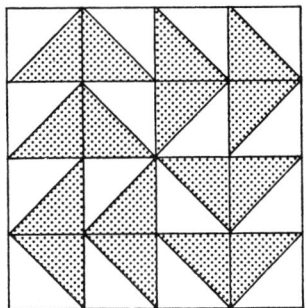

FRIEZE IN BLOCK PATTERN – UNDERSTANDING OF REPETITION PATTERN

Many unique patterns are possible when the grid is sized to fit the design. This frieze was extended by taping on another section of the 100-square grid. The importance of this design is in its repetition and harmony of color (Figure 1.21).

Figure 1.21

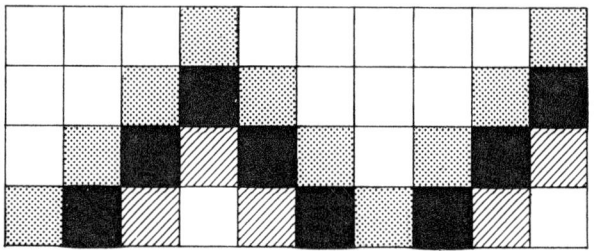

BORDER DESIGNS: INITIALS, MATH SYMBOLS, CROSS-STITCH – CONSERVATION OF AREA

Each repetition of the motif fills a certain area. Check the area involved and then block the motif to fit (Figure 1.22).

Figure 1.22

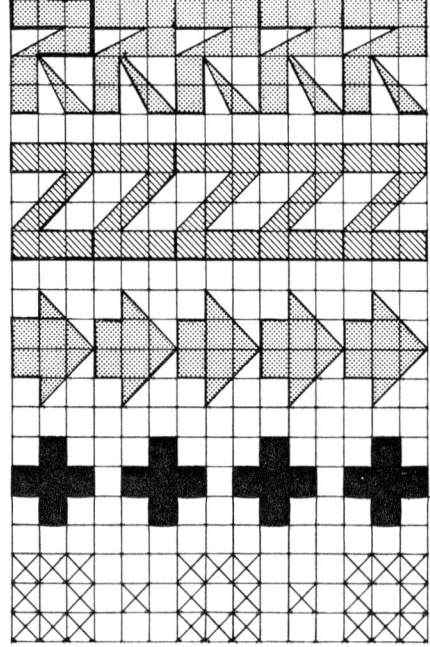

Math Activities—A Study in Design 25

BANDS FASHIONED OF DIAMONDS AND SPARKLE – CONTINUITY PATTERN

The only restriction was that the design be diamonds – the sparkle was a natural outcome. The end motif should be complete (Figure 1.23).

Figure 1.23

MODERN EFFECTS – FRACTIONAL PARTS – PERCENTS

A square matrix offers an inexhaustible field for experimenting. Freedom to explore and try out new ideas has special merit, even if the resultant art does not win a prize (Figure 1.24).

Mathematical awareness of fractional parts, percents, equivalence, square and rectangular regions, symmetry and nonsymmetrical form, decimal base, multiplication and division, arrays, and other algebraic and geometric abstractions are apparent when designs are arranged to satisfy the designer, and then discussed by himself and others.

Figure 1.24

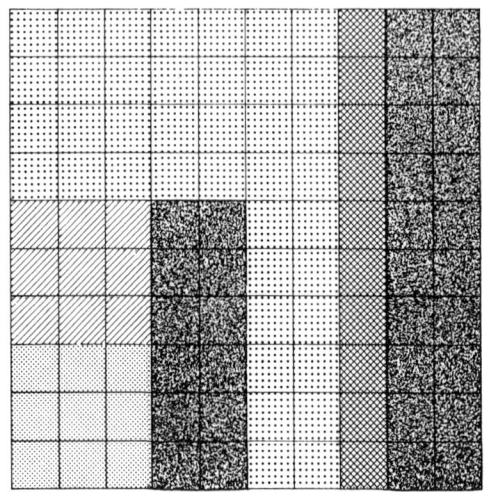

ROD ARRANGEMENTS – PERCEPTION

Pupils who have used colored rods in a mathematical laboratory situation, often like to fit them together in patterns that involve both size and color. Similar rod effects are created on the squared paper (Figure 1.25).

Figure 1.25

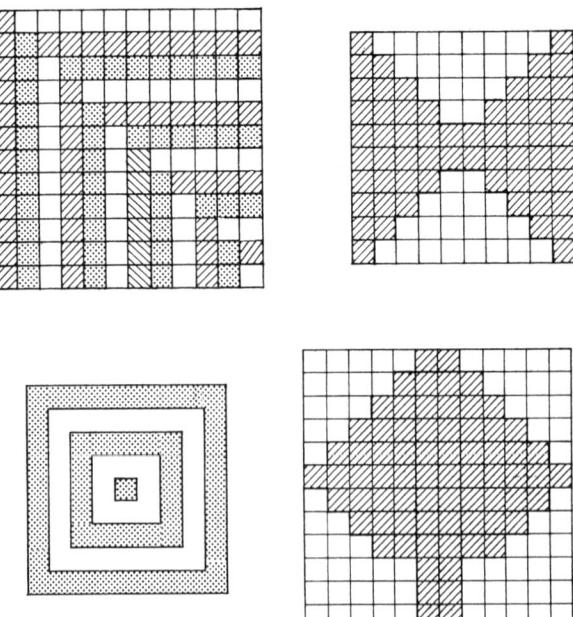

Math Activities–A Study in Design

COMBINING GEOMETRIC BLOCK DESIGNS AND NUMBER THEORY

These forms may be easily cut from the large 100-square and taped to the chalkboard. After the patterns have been determined and continued further, the blocks may be numbered. Discussion should prompt a name for each series and some important ideas in number theory.

ODD NUMBER SEQUENCE

Figure 1.26

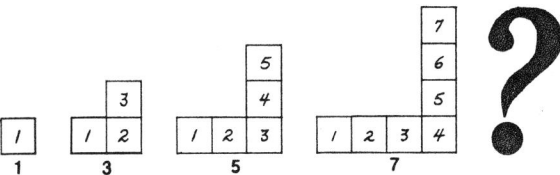

Joining the squares of each successive shape in Figure 1.26 results in each successive square shape in Figure 1.27. By this action we see that the addition of successive odd numbers produces successive square numbers.

$1 + 3 = 4 = 2^2$ $1 + 3 + 5 = 9 = 3^3$ $1 + 3 + 5 + 7 = 16 = 4^2$

The sum of the first 2 odd numbers is the square of 2; the sum of the first 3 odd numbers is the square of 3. What is the sum of the first 10 odd numbers? of the first n odd numbers?

Figure 1.27

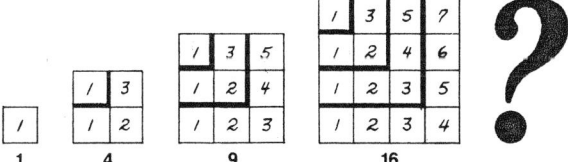

SQUARE NUMBER PROGRESSIONS (Figure 1.28)

Figure 1.28

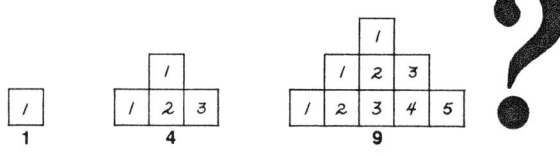

LINE SEGMENTS ON SQUARED PAPER – BASIC GEOMETRY

Short and long line segments and a creative spirit are the basic needs for these representative designs.

Line segments give a definite sense of movement to a design. We think of lightning striking from cloud to ground; of rain slashing through the sky; of a pathway cut of a maze; or of the course of an airplane.

In mathematics, vector and graphic geometries deal with movement, and quantities having both magnitude and direction (Figures 1.29, 1.30).

Figure 1.29

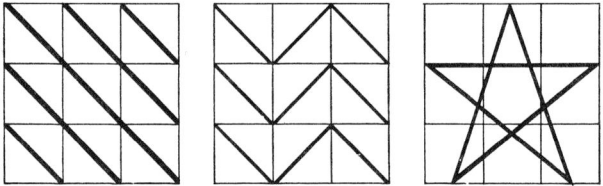

Figure 1.30

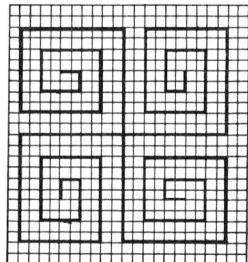

CURVED PATHS FROM LINE SEGMENTS – FORMING PARABOLAS

Convert the 100-square to a 20 x 20 grid by drawing horizontal and vertical lines with a fine point felt pen.

1. Number points around the sides as shown in Figure 1.31.

2. Working only in the A quadrant, draw lines in a one-to-one correspondence: 1 to 1; 2 to 2; 3 to 3; etc. Proceed in the same manner for each of the other quadrants, B, C, and D.

3. For the center design, number across and down the middle lines or axes. Working each quadrant separately, draw lines in a one-to-one correspondence.

4. Various colors may be used.

Figure 1.31

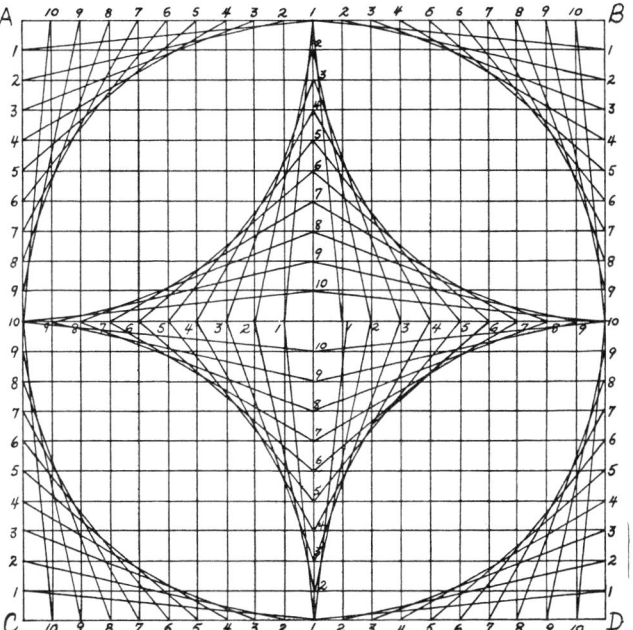

CIRCLE SYMMETRY – SKILL IN COMPASS USE

The lines and points on the squared paper enable pupils to judge whether or not they are using their compasses with precision, and to improve their skill in the use of this geometric tool. Only one row of squares is needed for the design in Figure 1.32.

Figure 1.32

The design in Figure 1.33 is produced by both curved paths or arcs, and line segments. Five rows of squares are needed for this design, and then the design motif is inverted or reflected in another five rows. How many axes of symmetry can you find?

Figure 1.33

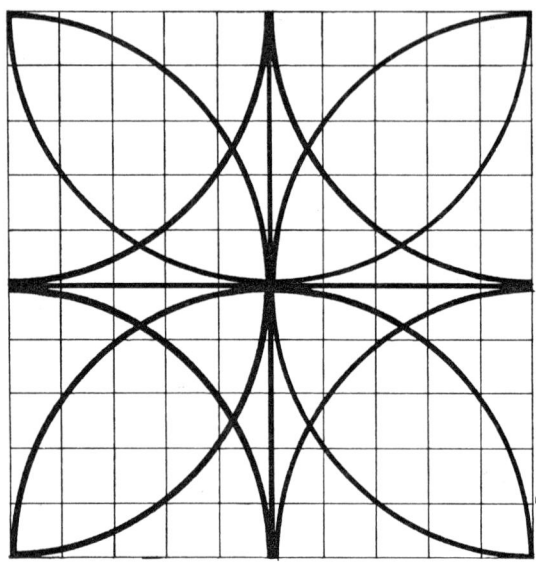

CREATE A FOLLOW-THE-NUMBER DESIGN – PRACTICE COUNTING ORDER

Older children create, outline, and number animal, toy, people, or flower pictures. Younger children love to play these follow-the-number games and learn counting order from them.

Directions: Use a black crayon. Start at 1 and draw along the line to 2. Next go to 3, and keep drawing to each numeral in order until 19 is reached. Color the interior of the closed curve with a brown crayon (Figure 1.34).

Figure 1.34

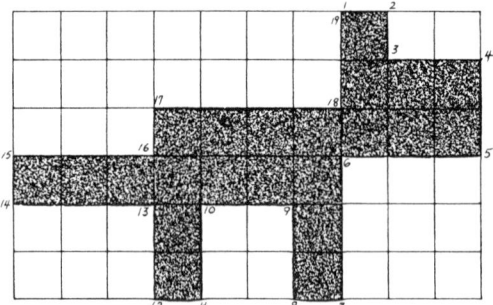

Math Activities–A Study in Design 31

Directions: You will draw a blue ? if you follow the lines from 1 to 16 Draw an eye. Write a word in the blank (Figure 1.35).

Figure 1.35

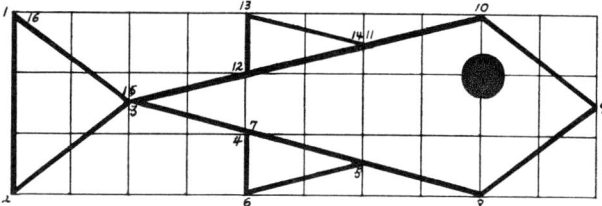

2.

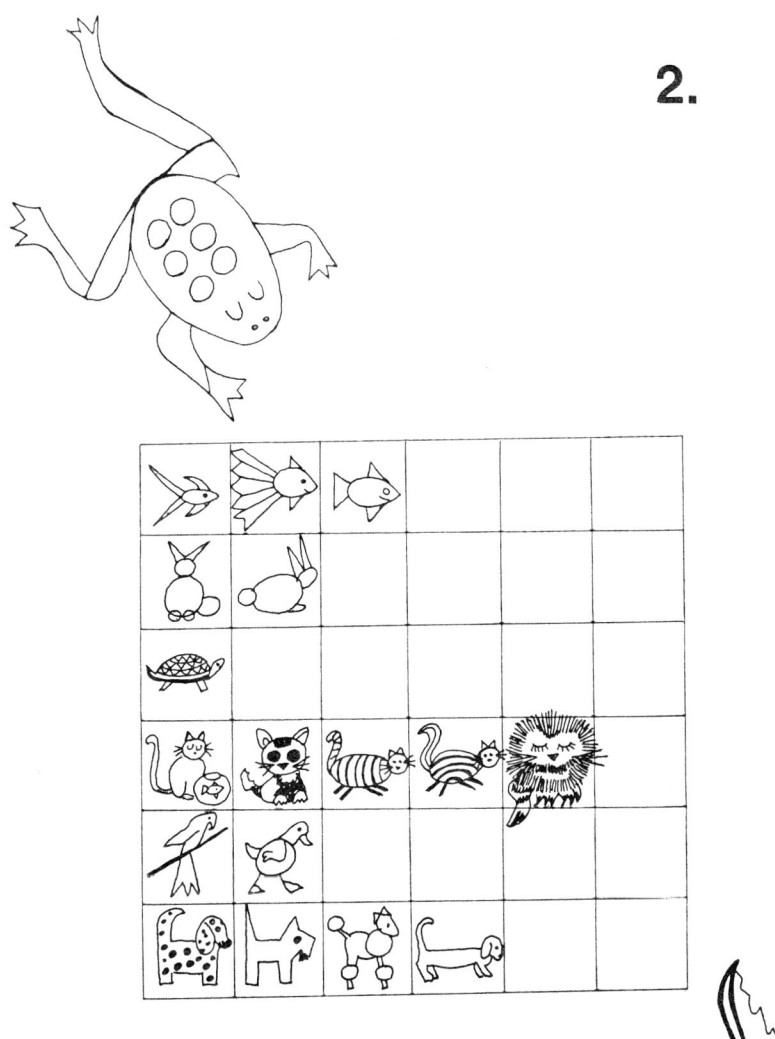

2

 **The Visual Impact of Graphs**

—information, organization, representation

To a kindergartener or other young child, a birthday is a very important occasion — all other holidays belong to everyone — but if a child was not born a twin, or on Father's birthday or Halloween, et cetera, his birthday is his alone. Besides being king for a day, he accomplishes an important thing in childhood — he gets a new age — one year older. He has added 1 to his number of years on earth, and numerically is "greater than" he was yesterday.

If you believe that the subject of graphs is too boring or too deep for most children, begin with the birthday graph. The theme is versatile — make one graph for the months in which they were born, another for the day numbers of their births, and if they know it, a third for the days of the week. Show the information by pictures, bars, tallies, or lines. Pasting on their own self-portraits makes the activity extra special for small children. Be sure to discuss the results. Children and other people are proud and conscious of their birthday statistics — as clearly indicated by the popularity of horoscopes.

The Visual Impact of Graphs

Great differences that show up in graphic material are also exciting, and great differences do show up beautifully in graphs. Compare the height of your school with that of the Empire State Building in New York City. Start at the bottom of the bulletin board and stack the squares upward, one square representing each story (about 12 feet). Build the school first and then the famous skyscraper. The Empire State Building is 1250 feet high and has 102 stories. The squares will reach the ceiling and then some. The children are impressed and eager to try new graphs. The teacher is eager to continue such an interesting subject with her class.

No need for a letdown. Try graphing Mt. Everest and the height of some landmark your children know. Even if they are from New York City and suggest the Empire State Building as a landmark, Mt. Everest's 29,028 feet is over twenty-three times as tall. Use ratio and proportion: if 10 squares represent the Empire State Building, then 230 are needed for Mt. Everest, and a comparable rate would be 5 to 115; 2½ to 57; 2 to 46; or 1 to 23. Since the mountain is measured from sea level, subtract the elevation of your area from that of Mt. Everest before presenting the relative sizes on a graph.

Since the visual impact of graphs is likely to be potent, they are recommended as motivational techniques for beginning lessons in many areas of study. A sample lesson on cotton production is given in detail in this chapter. Here the children provide the questions and the purpose of the lesson by studying graphs provided by the teacher. Here they also make use of graphs to discover the answers.

Although graphs appear to be simple charts, practice is needed in order to read, interpret and fully comprehend the diagrammatic information shown. Collecting and organizing the data necessary for the construction of these factual drawings also requires special skill. Details must be noted. For example, numerical quantities shown on a graph are usually not exact but "rounded numbers", which in many cases are the most meaningful. Some graphs use widely different scales for the vertical and horizontal reference lines. For reasons such as these, and because of the ever increasing amount of information shown in the form of graphs at the present time, exploration and a thorough study of graphs are recommended for all grade levels.

Line graphs suggest the graphing of solutions sets, linear equations, slopes and functions on a coordinate grid. See Chapter 9 for coordinate graphs.

Figure 2.1

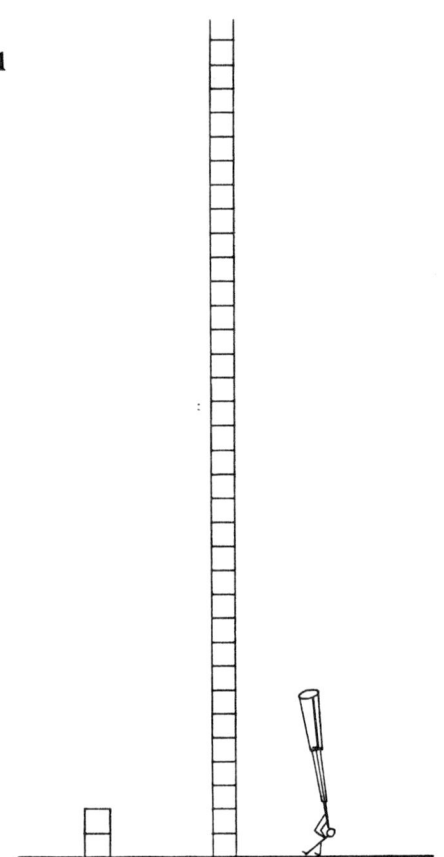

INVITE ACTION WITH A BIRTHDAY GRAPH

The grid is an important aid for organizing numerical data. First, put data in order and determine the two associated facts that will be shown on the graph. These facts which make up an "ordered pair" could be the name of a month, Jan., and the number of children in the class who have birthdays in that month, 4. Data may be recorded on the hundred-square grid by using tally-marks, pictures, individual squares as units of a bar, or by lines connecting points (Figure 2.2).

Figure 2.2

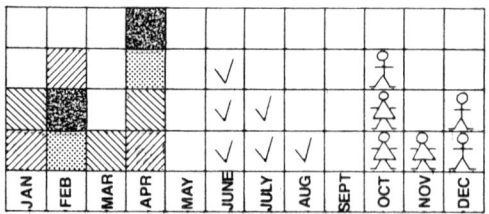

The Visual Impact of Graphs

The "birthday" graph is a class project that each child helps to construct. It may be used in kindergarten or primary classes. Prepare a 12 by 12 grid for the 12 months of the year. Although three ways of recording data are illustrated on the sample graph above, only one type should be used for any one graph.

1. *Bar Graph* — Cut a block for each child to color and paste on the grid to show his birthday month.

2. *Tally Graph* — A check mark is made by each child to indicate his month.

3. *Pictograph* — Have each child draw his own picture on a square and paste it on the graph.

The graph may be shown in either a horizontal or vertical position.

Check the accuracy by having the children act out the graph. January's children stand in the first row, February's in the next row, March's children in the next, and so forth. The children themselves form a live graph to match the paper version. Each month's children count off to verify the paper birthday graph.

GRAPHS FOR MEANING OF NUMBER

A graph is a beneficial visual aid for developing many arithmetic concepts and relationships. Discussion of what the graph represents, of *all* the information it gives, of how this information is organized and may be further utilized, is as essential to the lesson as the construction of the graph itself.

The graph (Figure 2.3) depicts 1 as the first odd whole number; the next odd number, 3, as 1 and 2 more; and each succeeding odd number as 2 more than the one preceding. Commuting this pattern allows us to write the formula for odd numbers as $2n + 1$ where n is any whole number. This formula suggests that 11, 13, 15, 17 and 19 are the teen odd numbers. This continuing pattern awakens the awareness that the digit in the ones place is the clue for detecting and naming an odd number. Does the clue imply that 10, 30, 50, 70 and 90 are odd numbers? Why, or why not? Could these be named the odd tens?

Figure 2.3

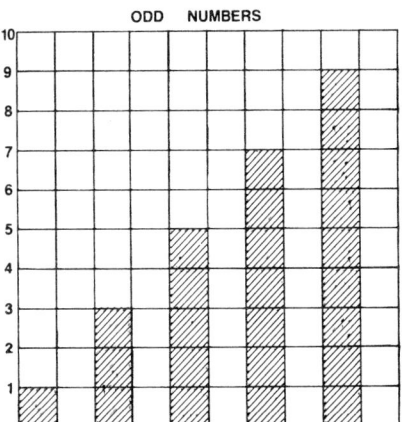

GRAPHS FOR PRACTICING FACTS

Bicycle *wheels* and the second table naturally go together. Bi- is a prefix meaning two and bicycles help children think by twos. Counting by twos from zero gives the multiples of 2. These numbers are the products when 2 is used as a factor. The bicycle technique rates as the adroit strategem, but socks, gloves, earmuffs, boots, and other pairs may be employed for further practice (Figure 2.4).

Use tricycles, clovers, and triangles for graphing the third table. Letting children pick up toy cars (4 wheels) as they learn their fourth table, and then graphing this action is a highly successful experience. Try nickels for the fifth table.

Figure 2.4

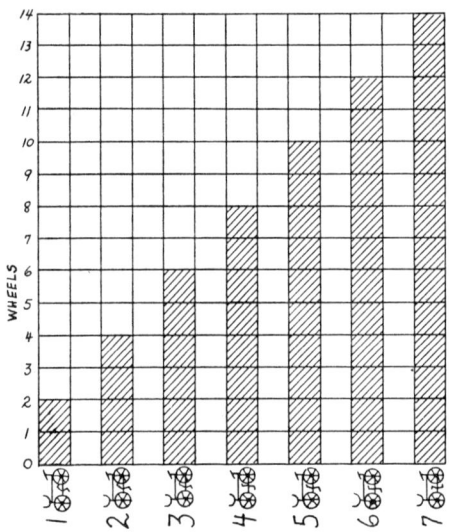

GRAPHS FOR KEEPING RECORDS

Individual graphs showing results of mathematics, spelling, or other tests, give significant information to the pupil involved and to his parents and teacher. Even a quick stop and look at the diagrammatic data is more likely to make a pupil realize and restudy than a thousand words. Further interpreting and analysis are advantageous.

The 100-square allows raw scores to be quickly changed to percentages. To construct the graph in Figure 2.5: count and number by tens up the left side for the number of problems solved correctly. For example, color up seven squares for a score of seven out of ten correct. This shows 70 and means 70% of the problems are correct. At the completion of ten tests the child has a record of his achievement. This type of graph may be used successfully with primary children, who develop a feeling for percentage although no formal explanation is given in these early grades.

Figure 2.5

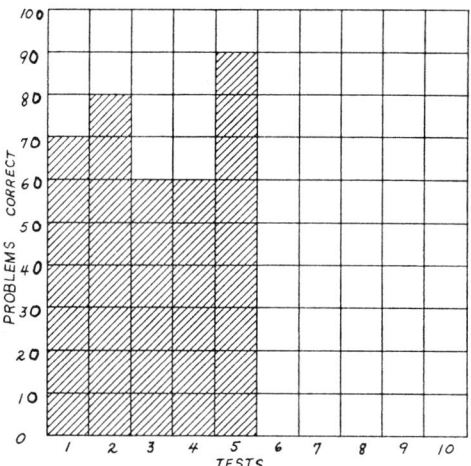

CLASS ACHIEVEMENT RECORD

A vertical bar graph is suitable for consolidating class achievement for a particular test. (See Figure 2.6.) If two children received a score of 10, fill two blocks in Column 10; the number having 9 correct is shown in the ninth column; and similar scoring is continued until all the scores are properly distributed. This distribution of scores may simulate a normal bell curve, with the highest frequencies in the middle or average range, and the few scores that were very high or very low on the extremities of the graph. (See Figure 2.7.)

If the distribution of scores produces a graph skewed to the left, it will have a long left tail, and this will imply that the test was too easy or that the class is exceptional. (See Figure 2.8.) A graph skewed to the right will have a long right tail and will mean that the test was unduly difficult.

The average known as the *mode*, or the most frequent score, is represented by the tallest bar. In Figure 2.6, the mode is 7. Five students achieved this score. The middle number in the correctness scale is the average known as the *median*. For the 27 recorded scores in Figure 2.6, the 14th score is the median and it is 6. To find the *arithmetic mean*, or simply the mean, total all the scores and divide by the total number of scores entered. For Figure 2.6: (1 x 1) + (3 x 2) + (4 x 3) + (5 x 4) + (6 x 4) + (7 x 5) + (8 x 4) + (9 x 2) + (10 x 2) = 168. 168 ÷ 27 = 6. The mean is 6. The average person means the mean when he speaks of an average.

Figure 2.6

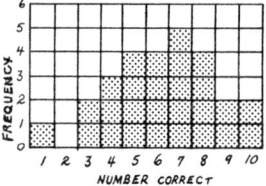

Since a spelling test is actually given to the pupils to study before they take it, should the scores produce a graph skewed to the left? What type of graph would be formed by the ages of the pupils in the class?

Figure 2.7   **Figure 2.8**

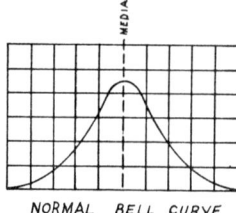 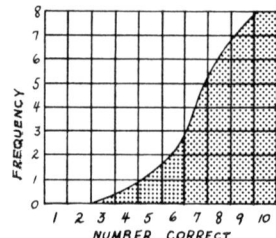

INDIVIDUAL PUPIL'S ITEM ANALYSIS GRAPH

In Figure 2.9, the star * indicates the number of the particular problem the pupil worked incorrectly. A dot • is used to show the problems that were *not tried*, if this information is desired.

An item analysis graph kept by each pupil allows consolidation of information by the teacher and helps in planning necessary reteaching details. It gives knowledge to individual pupils of needed practice and restudy, and supplies interested parents, who wish to assist their children, with information about the area where help is needed.

Figure 2.9

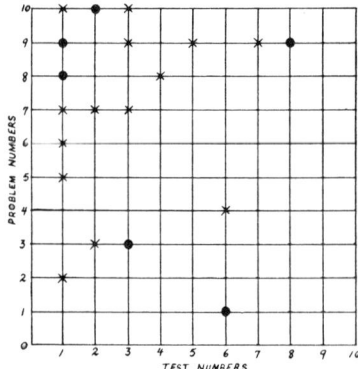

The Visual Impact of Graphs   *41*

GRAPHS FOR INTRODUCING LESSONS – STORY PROBLEM

This graph has been used many times to introduce a cotton production lesson in a social studies situation. It has also been used equally as often in a mathematics situation to introduce graphs. It has never failed to be an interesting beginning technique.

When students are asked to study the pictograph showing the growth of cotton production in the United States from 1790 to 1796, they invariably ask, "Why was there such an increase in the amount of cotton raised in the years from 1792 to 1794?" Having established a purpose, the students approach their reference reading with curiosity and incentive, and learn that Eli Whitney invented the cotton gin in 1793. This research, in turn, leads them on to related subjects on cotton growth and climate requirements; to the study of temperature and rainfall graphs; to the construction of graphs showing the leading cotton production states of the United States, and to other graphs showing the most productive cotton countries of the world.

Further study leads to other purposeful questions that include: How large is a bale? Does the pictograph represent exact or rounded figures? When was the next sharp change in the production statistics of cotton? (Figure 2.10)

Figure 2.10

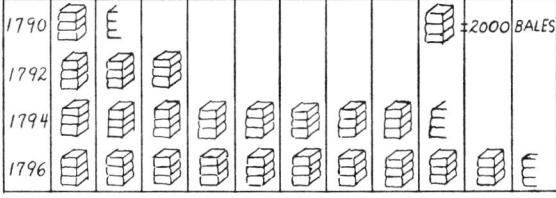

COMBINATION TEMPERATURE-RAINFALL GRAPHS – STORY PROBLEM

Line graphs represent continuous data, while bar graphs depict information that is not continuous. Figures 2.11 through 2.15 are combination graphs, the lines of each graph showing continuous temperature data, and the bars representing monthly rainfall in total amount of inches. Each graph gives these average monthly climatic conditions for specific areas, and are used in the lesson on cotton production.

Students learn the following information from their research: Cotton grows best in a very warm climate and needs about 186 growing days without frost. Graphs of cotton growing areas show the average temperature for every month to be above the freezing point. There must be sufficient rain in the spring and early summer when the young plants are growing, but too much rain in the late summer and fall would ruin the mature, bursting cotton bolls. Given this information, students are asked to judge where they think cotton might be grown commercially. Names of places are covered.

Figure 2.11

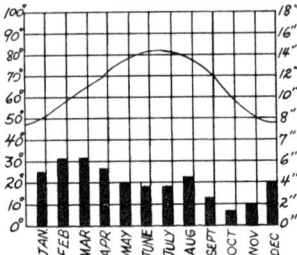

Yes, the temperature is right, and the rainfall is well distributed, with very little precipitation in the months of September and October when cotton is harvested. The place is Montgomery, Alabama.

Figure 2.12

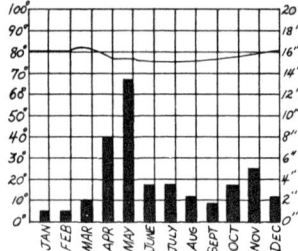

Yes, with temperature that is tropical all year, and rainfall that is light in the harvesting season, cotton is successfully raised in this part of Africa. It's Mombasa, Kenya.

Figure 2.13

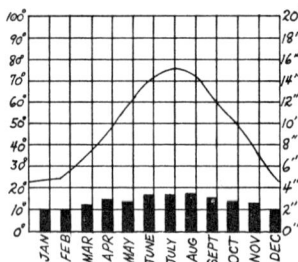

No, rainfall is sufficient, but the growing season is too short. This climatic graph is for a ten-year period at Detroit, Michigan.

The Visual Impact of Graphs

Figure 2.14

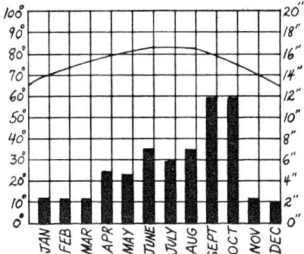

No, good cotton-growing temperature here, but the hurricane season comes at the time when the fluffy cotton bolls are open and is one of the reasons very little cotton is raised in Florida. This is Miami.

Figure 2.15

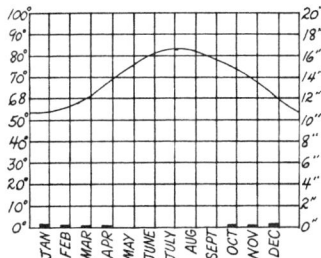

Yes, the temperature is tropical, but the rainfall is insufficient. Yet this is the climatic graph of one of the great cotton producing countries of the world. How is this possible? (Cairo, Egypt.)

GRAPHS FOR COMPARISON

The average number of bales of cotton harvested in the year 1945 is compared with that raised in 1965 for four leading cotton producing countries. The double bar graph (Figure 2.16) pictures the special comparison between the years given, and at the same time, the changes in production and leading production order. The figures in the table were rounded to the nearest 500,000 for determining the allotment of squares on the graph.

TABLE GIVING NUMBER OF BALES

Country	1945	1965	1945	1965
United States	13,800,000	14,956,000	~ 14,000,000	15,000,000
India	4,431,000	4,500,000	~ 4,500,000	4,500,000
China	2,897,000	5,700,000	~ 3,000,000	5,500,000
Russia	2,063,000	8,700,000	~ 2,000,000	8,500,000

Figure 2.16

UNITED STATES										
1945	░	░	░	░	░	░	░	░	░	░
1965	▨	▨	▨	▨	▨	▨	▨	▨	▨	▨
INDIA										
1945	░	░								
1965	▨	▨	▨							
CHINA										
1945	░	░								
1965	▨	▨	▨							
RUSSIA										
1945	░									
1965	▨	▨	▨	▨						

Several 100-square grids are taped together for this class project. Each shaded square represents 1,000,000 bales.

EXPERIMENTING WITH LINE GRAPHS

Broken-Line Graph

A graph showing a sharp rise and fall pattern is called a broken-line graph. Figure 2.17 showing postage rates has a different type of broken line. Letters that weigh exact ounces would be shown to have many different postage rates if the line segments were connected to form a continuous line.

Figure 2.17

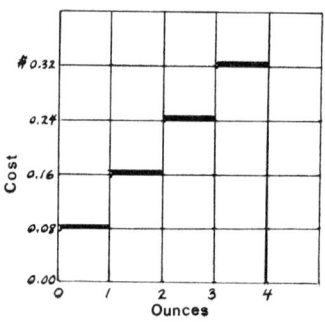

The construction of a line graph is very similar to constructing a graph of a solution set on a coordinate grid. Each point is named by an ordered pair, and located at the intersection of horizontal and vertical lines measured from an origin point.

Line graphs may be used as pre-coordinate activity and lead directly to a study of the graphing of equations and functions on a coordinate grid. See Coordinates, Chapter 9.

Batter-Graph

We assumed that champion batters should get one hit every three times at

The Visual Impact of Graphs

bat, and tested this hypothesis by constructing a batter-graph showing the records of 22 champion batters from 1960 through 1970:

American League

			AB	H	Av
A	60	Runnels (Bos.)	525	168	.320
B	61	Cash (Det.)	535	193	.361
C	62	Runnels (Bos.)	562	183	.326
D	63	Yastrzemski (Bos.)	570	183	.321
E	64	Oliva (Minn.)	672	217	.323
F	65	Oliva (Minn.)	576	185	.321
G	66	Robinson (Balt.)	576	182	.316
H	67	Yastrzemski (Bos.)	579	189	.326
I	68	Yastrzemski (Bos.)	531	160	.301
J	69	Carew (Minn.)	458	152	.332
K	70	Johnson (Cal.)	615	202	328

National League

L	60	Groat (Pitts.)	573	186	.325
M	61	Clemente (Pitts.)	572	201	.351
N	62	Davis (L.A.)	665	230	.346
O	63	Davis (L.A.)	555	181	.326
P	64	Clemente (Pitts.)	622	211	.339
Q	65	Clemente (Pitts.)	589	194	.329
R	66	Alou (Pitts.)	535	183	.342
S	67	Clemente (Pitts.)	585	209	.357
T	68	Rose (Cinn.)	626	210	.335
U	69	Rose (Cinn.)	627	218	.348
V	70	Carty (Atla.)	478	175	.366

Figure 2.18

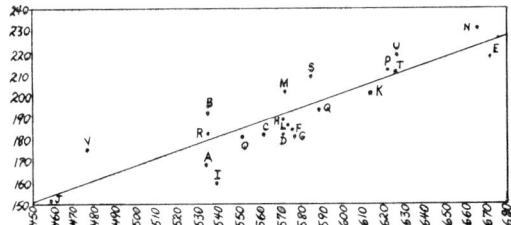

Key:
1. The line is a graph of the ratio 1/3 or .333.
2. For convenience, the axes are numbered by tens to include the lowest and highest scores. Times At Bat (AB) is shown on the horizontal axis and Hits (H) on the vertical axis.

3. The points are located by the ordered pairs (AB, H) and are letter coded to the players in the list.

Percentages are the ratio of actual Hits to actual Times At Bat compared to potential Hits (x) to 1000 Times At Bat. $\frac{H}{AB} = \frac{x}{100}$

Students are to estimate players' percentages by their position on the graph in relation to the line and then compute the actual percentages.

GRAPHS SUITABLE FOR VARIOUS GRADE LEVELS

All teachers know that the lessons that pay the highest dividends in effort put forth and achievement scored are those that truly interest the children. The titles listed below may or may not appeal to your classes, but they will surely suggest data that will rouse some student to do the research and work necessary to produce a graph intriguing to himself; and when he interprets the specialized knowledge it contains, it becomes a worthwhile lesson for the entire class.

Primary Grades

1. Long socks, short socks, tights

2. Pets at home

3. Number of children in the family

4. Make of car owned by family

5. Favorite color

6. Number living on each street

7. Absences for each day of the week

Upper Grades

1. Population of nearby towns; of largest cities in the world; of nearby schools

2. Production of rubber, wheat, mink, sugar, TV sets, telephones

3. Size of planets, continents, rivers, mountains, lakes, oceans, states, buildings

4. Sport scores: Little League, bowling

5. Favorite pie, fruit, soft drink, snack

6. Disliked vegetable

7. Test scores

8. Child's own growth for a year

9. Rainfall of tropical rain forests, deserts, your town

Junior High Grades

1. Football statistics, bowling, champion batters in the big leagues, track records

2. Favorite hobbies

3. Favorite recording, rock group, pop singer, classic, actor, TV program

4. Favorite sandwich, dessert, candy bar

5. Outside jobs

6. Size of airplanes, space ships, rockets, boats

7. Favorite active sport and favorite spectator sport

8. Graphs of probability – coin tossing, drawing colored beads from box (See chapter on probability.)

9. Stock fluctuations

3.

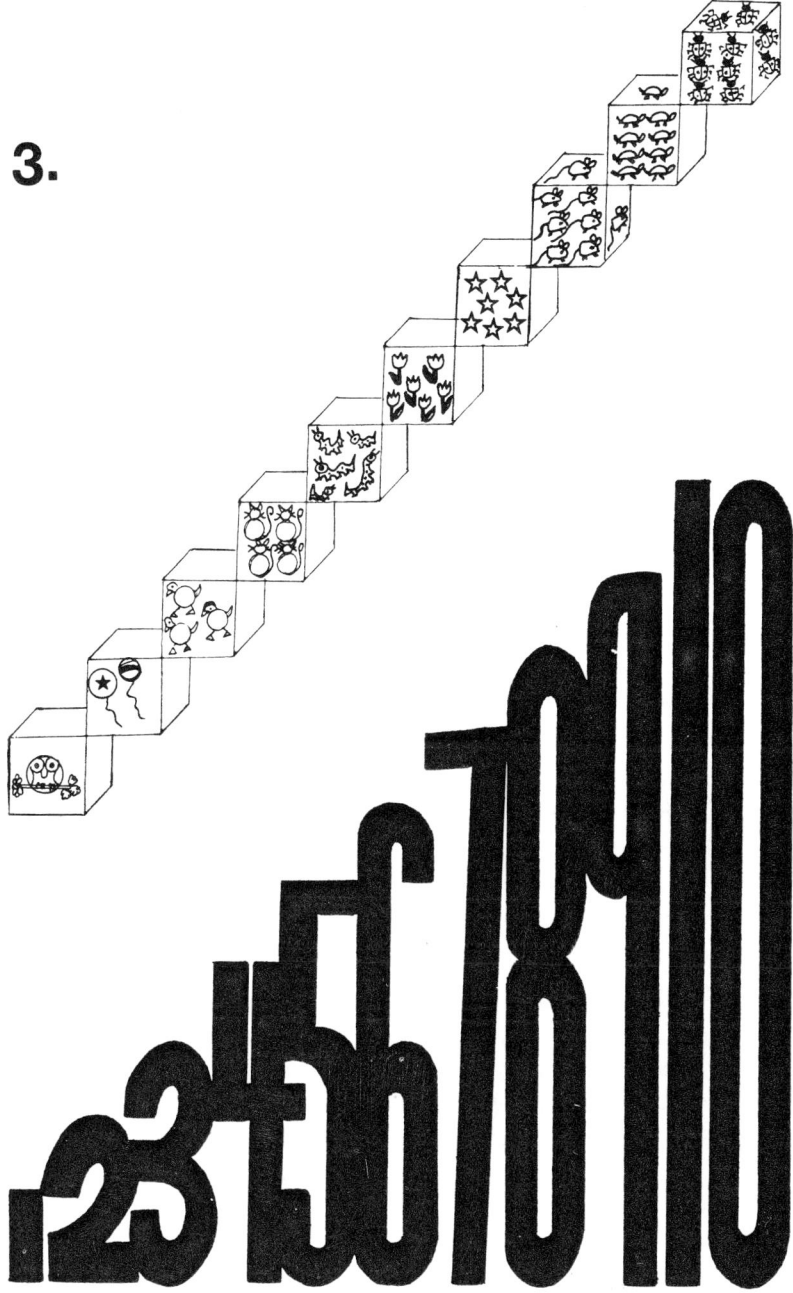

3

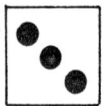 **Understanding Number**

—symbols have a high level of abstraction — first build a high level of understanding

To play a certain guessing game, you are told to think of a number, add, subtract, multiply, or divide other numbers to or from it, and then to tell or write the result. When the idea was in your mind and you were carrying out the orders, you were involved with and operating on *numbers*. When you told or wrote your answer, you were involved with *numerals*, which are the words and symbols used to express the number idea that was in your mind.

When beginning the study of number, an understanding of the number system and the numeration system go hand in hand. When "5" is written, or "five" is said, the child should think of a set with a "fiveness" quantity, and similarly, when shown any set with a "fiveness" quantity, he should be able to identify it by word or written symbol.

Our decimal number system is, as the name implies, a base ten system. Operations of counting, adding, subtracting, multiplying, dividing, doubling, factoring, and so forth, are performed on numbers.

Understanding Number Symbols

Our numeration system essentially consists of: (1) a set of ten symbols [0, 1, 2, 3, 4, 5, 6, 7, 8, 9], and (2) a place value rule for forming all the numerals greater than 9, and with the use of the decimal point, the decimal fractional numerals.

These basic mathematical ideas establish a foundation for all work with number, and indicate that the beginner's program should include experiences that help children to: (1) develop a concept of number, both in a cardinal (how many) and an ordinal (which one) sense; and (2) learn how to read and write the numerals, using place value with understanding.

The number symbols, with their high level of abstraction, are to the primary child as a sheet of music is to a person who cannot read notes. The primary teacher, realizing this, does not restrict the teaching of the symbols and place value entirely to "what's in the book", but concretely associates numbers and objects, and searches for suggestions of numerous types to meet the diverse needs of the class, and to bridge the gap between the concrete and abstract. The work of laying a good foundation in these essential areas will be appreciated by all subsequent mathematics teachers.

One-to-one correspondence between members of sets emphasizes equivalent and nonequivalent sets, and is an important technique for building an understanding of number and the resultant numerals. The act of matching sets points out the order principle. For any two numbers, one and only one of the following conditions is true: the first number is (1) equal to; (2) less than, or (3) greater than the second number.

The 100-square is especially suitable for these and many other number activities. The 2-dimensional spaces are convenient whether a child forms a larger set by coloring one more square, or writes within the squares as he practices his figure 8's. This chapter provides many more how-to-do-it helps that implement understanding of number and numerals.

Figure 3.1

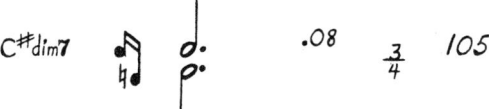

EQUIVALENT AND NONEQUIVALENT SETS

Comparison of two sets by a one-to-one matching, first using concrete objects and then semiconcrete pictures, colors and shapes, is a suggested activity for developing number understanding.

Use Figure 3.2 and have individual children draw lines to show the mapping of the girls and swings. Ask specific questions: Does every girl have a swing? Does every swing have a girl? Are there more girls than swings? Are there any swings left over? Are there the same number of swings as girls?

Figure 3.2

The sets in Figure 3.2 are equivalent. They have the same number of members.

Figure 3.3

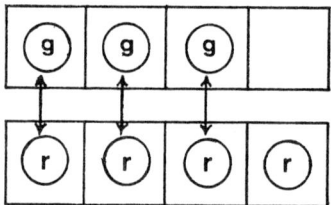

The sets in Figure 3.3 are not equivalent. The green set has *fewer* members. The *red* set has *more* members.

Figure 3.4

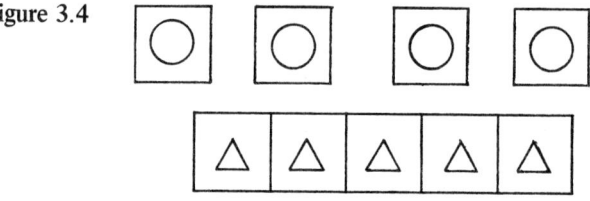

Young children commonly believe that a set that occupies more space than another set has more members. Matching is an effective way of showing that 5 is greater than 4, and 4 is less than 5.

$$5 > 4 \text{ and } 4 < 5$$

"Count 4 *before* 5. Count 5 *after* 4."

COUNTING AND NUMBER ORDER

These "square" activities are designed to practice the counting order of the

Understanding Number Symbols

cardinal numbers, but at the same time they give meaning to the ordinal number concept.

Cut sets of large size squares and tape them in a random arrangement on the chalkboard or magnetic board. A child is directed to count as he flies a cardboard bird from the shortest post to a post that has one more square, until the bird is on the highest post. The bird flies back down touching each post in decreasing order.

Figure 3.5

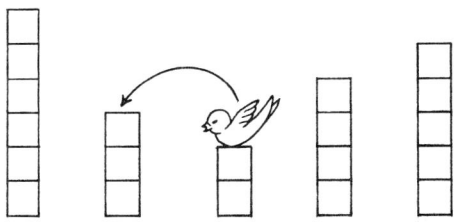

Jiminy Cricket (or Jumping Frog) hops on these squares in a 1-more pattern.

Figure 3.6

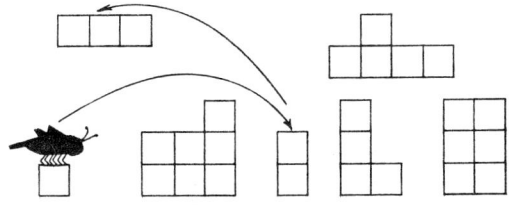

NUMBER SYMBOLS AND PICTURES

Count the number of pictures in each row in Figure 3.7. Write the numeral in the next square.

Figure 3.7

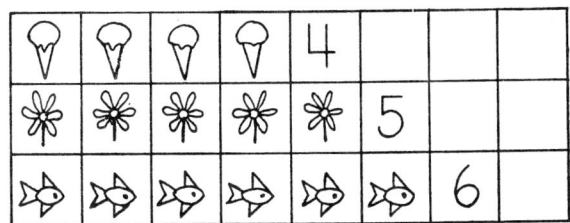

Draw the correct number of pictures in each row in Figure 3.8

Figure 3.8

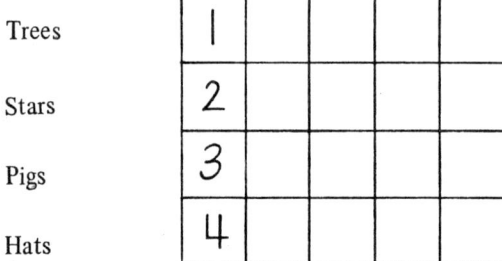

Children are given a two-square card for each number as it is developed. They write the numeral and draw the dots. To show the relationship of the numbers to each other, they compare the number of dots on the cards and arrange them in counting order. See Figure 3.9.

Figure 3.9

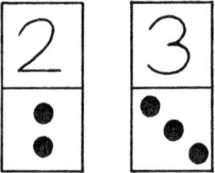

NUMBER SYMBOLS AND DOTS

When the abstract numerals are first introduced they are matched to a set of objects, next to a set of pictures, and then to a set of dots (Figure 3.10).

Figure 3.10

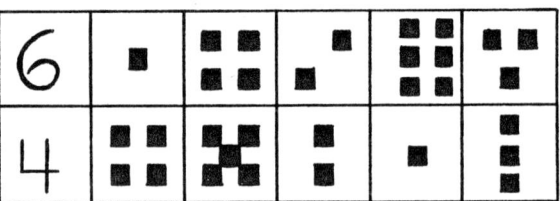

NUMBER SYMBOLS, NUMBER WORDS, AND SQUARES

In Figure 3.11, the numerals tell how many squares to color in each row.

Understanding Number Symbols

Figure 3.11

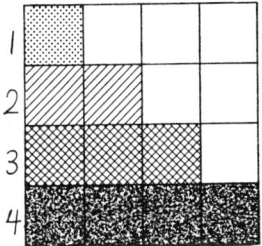

In figure 3.12, color one dot on the first cross on Line 1. The numeral tells how many dots to color on each line.

Figure 3.12

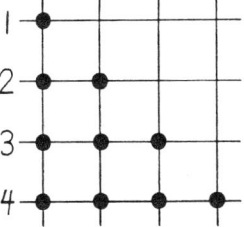

In Figure 3.13, color one dot on the bottom cross in Column 1. The numeral tells how many dots to color in each column.

Figure 3.13

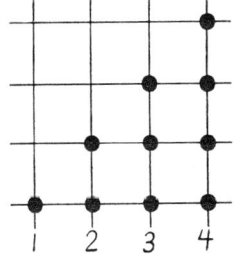

The number names tell how many squares to color in Figure 3.14.

Figure 3.14

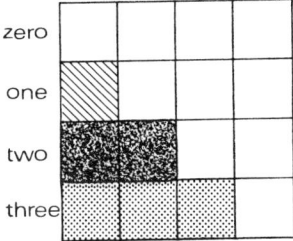

After order is firmly established, each exercise may be arranged in a random fashion. For quick review, use counters instead of coloring dots.

PRACTICE YOUR FIGURE 8'S

These individual practice cards, when laminated, may be written on with grease pencil or pointed crayon and wiped off for reuse.

The first numeral is a touch-me numeral cut from sensitized backing. The x in each square shows where to begin the numeral.

Tell a trick about each numeral. For the 8: "Make a road that looks like an S, cross the road and come back home (Figure 3.15)."

Figure 3.15

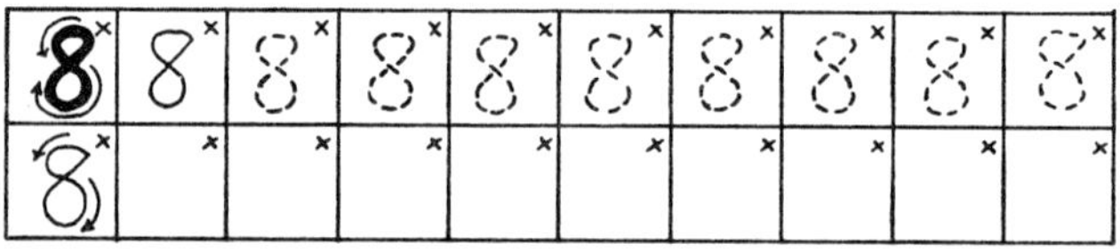

Begin at the dots and follow the arrows.

Figure 3.16

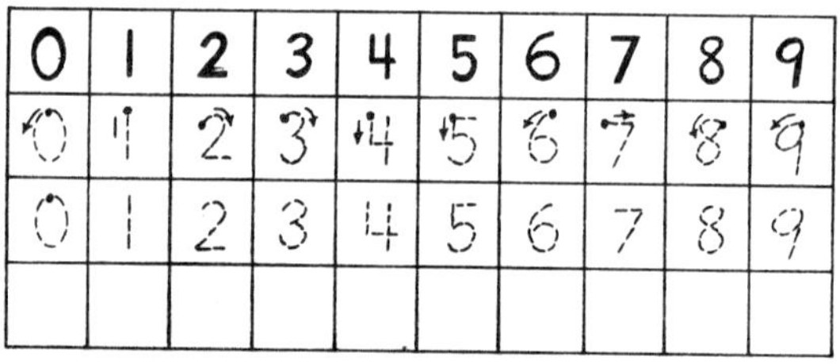

EVER-READY BULLETIN BOARD
NUMBER PRACTICE CHART (Figure 3.17)

For many weeks, this primary chart will provide a quick daily review. Ask questions about counting, tens, more, fewer, before, after, equivalent sets, nonequivalent sets, comparison of sets, how many more to make a ten, ordinal number concept, addition, subtraction, and so forth. Even when children are lined up, waiting for the dismissal bell – don't waste time – ask questions:

Could every rabbit have an umbrella? (equivalent sets).
Could every rabbit have a star? (nonequivalent sets).
How many flags are there? (cardinal number)
Point to the first boat. (ordinal number).
Count the stars. Find a set that has two more members. (Comparison subtraction).
Join the set of flags and the set of circles, and tell how many altogether. (Adding tens).

Understanding Number Symbols

Join the set of circles and the set of triangles. How many altogether? (Adding 10 and 9 to a teen sum).

Do you know a way of finding out how many triangles there are without counting? (Subtract 10 − 1 = 9)

How many triangles would there be if there were 3 more? (Regrouping).

Fill in: Leaves: 5 + □ = 10; Umbrellas: 6 + □ = 10; Flags: 10 + □ = 10; Boats: 2 + □ = 10 (Subtracting from 10)

Figure 3.17

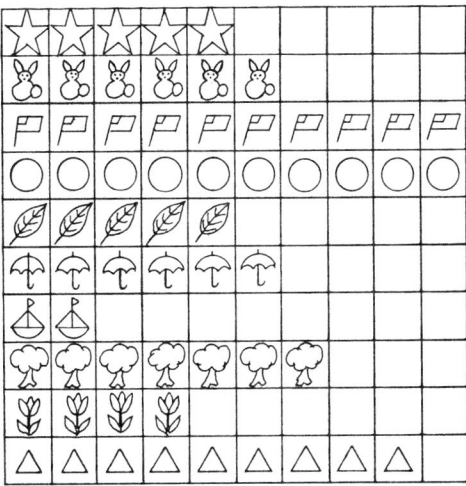

PLACE VALUE – BUILDING A TEN

Individual children tape paper disks in each square as the class counts out and builds a set of "ten."

TEN'S SYSTEM

Figure 3.18

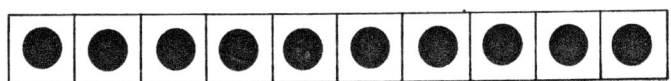

1 Ten

When all have agreed that the ten set is correct in number, it never has to be counted again. Whenever it is referred to, the children simply say, "Ten." As other "ten" sets are built, these sets and the extra "ones" are recorded in a table See Figure 3.19.

Figure 3.19

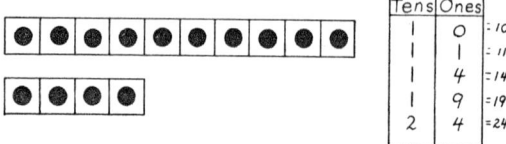

Later, each set is turned over and marked: 1 Ten, and is understood to represent the ten disks.

Figure 3.20

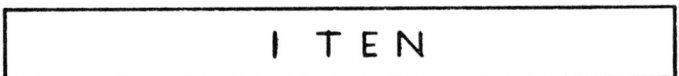

COUNT 1,000 SQUARES BY 100'S

Tape ten 100-squares together. This makes an impression and 1,000 becomes a large number to primary children.

Also let a large square represent 1,000; a medium size stand for 100; small for 10, and very small for 1. Figure 3.21 shows 1,351.

Figure 3.21

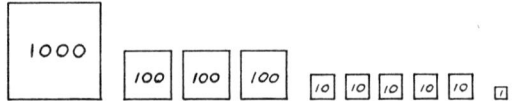

ARROW MOVES — A PRE-NUMBER LINE BULLETIN BOARD CHART

The fingers walk along the row of pictures at the top of the chart. The other rows contain seven problems. The first picture in each problem tells where the finger *starts* to walk and the arrows show how many *steps* to take and the direction in which to go. The child tells on which picture he *stops*. Sometimes the *stop* picture is given, and the child must tell the number of arrows needed and the direction (Figure 3.22).

Understanding Number Symbols

Figure 3.22

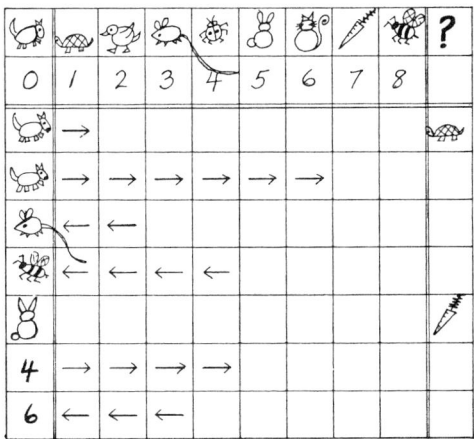

YOUR FINGERS DO THE WALKING

Along A Short Road
Figure 3.23

Up A High Hill And Down
Figure 3.24

Up A Tall Ladder And Down Again
Figure 3.26

Up And Down The Stairs
Figure 3.25

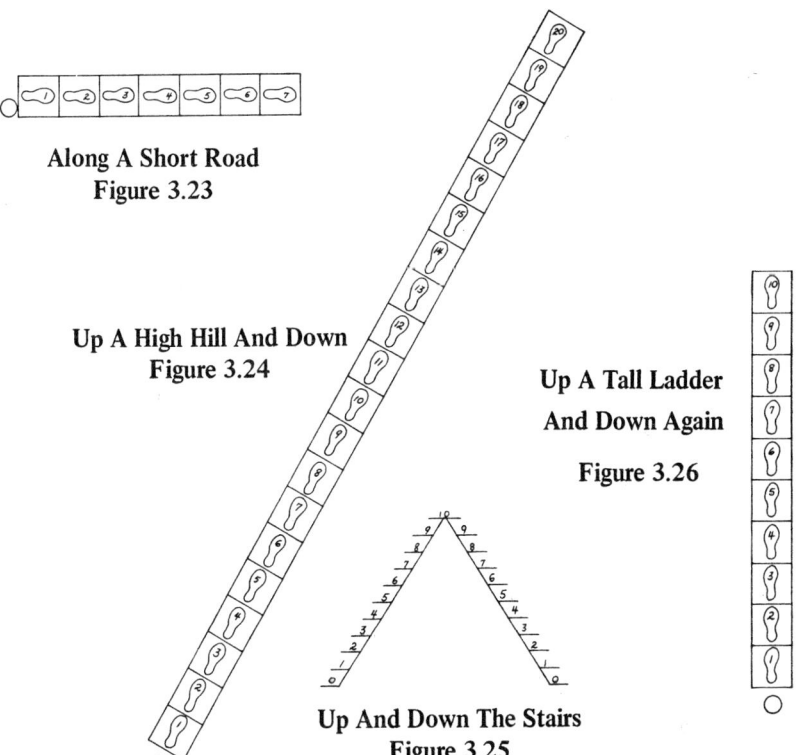

COUNTING ROWS AND NUMBER LINES

Figure 3.27 shows an individual counting row. Move from square to square and count the squares.

Figure 3.27

Figure 3.28 shows an individual number line. Move from point to point, but count the *intervals between the points*. To help children count correctly on a number line, move a cardboard frog along the line. They count only when the frog finishes each jump.

Figure 3.28

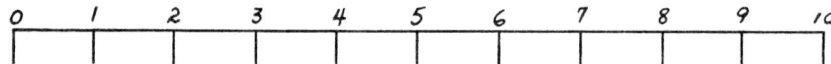

MARCH LEFT, RIGHT, UP AND DOWN

Give directions as shown in Figure 3.29. Write in the number that locates your stopping place.

Figure 3.29

Start	Go Right	Stop
0	2	2
6	2	
8	1	

Start	Go Left	Stop
8	2	6
6	4	
7	1	

Try a vertical number line (Figure 3.30) and directions that say: "Start Go Up Stop" and "Start Go Down Stop." Let the children have a chance to write the orders.

Understanding Number Symbols

Figure 3.30

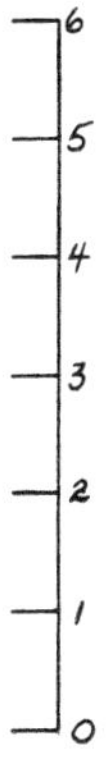

(vertical number line from 0 to 6)

NUMBER LINES FROM SQUARED PAPER

The points on squared paper are evenly spaced, so the 100-square is instant number line paper. Any type of number line can be fashioned by cutting the squares through the center and numbering appropriately.

A long number line that reaches all the way across the chalkboard is a wonderfully useful device for every mathematics room. Laminate it, and tape it to the top of the board if you teach older children, or along the bottom for small tots, so that loops or arcs may be drawn and easily erased.

Figure 3.31

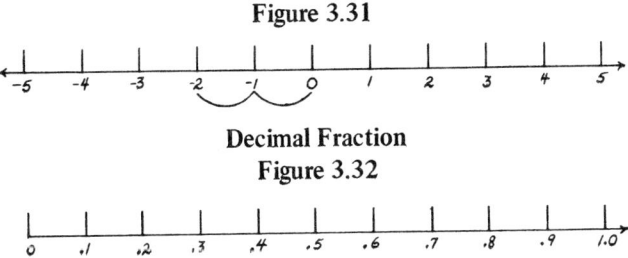

Decimal Fraction
Figure 3.32

An equivalence chart may be formed by placing number lines scaled in different fractional parts above each other.

Fractions
Figure 3.33

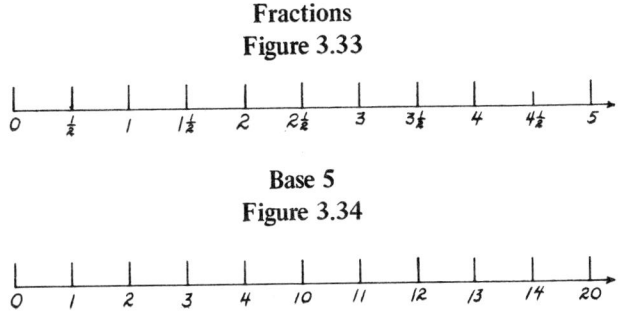

Base 5
Figure 3.34

Roman Numerals
Figure 3.35

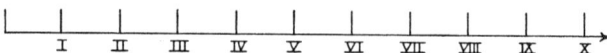

Rounding Numbers — **an Approximation Exercise that helps with Estimation.** For rounding to the nearest 100, (or 10,1000, etc.) use only the subset needed. 48 is not half-way to 100 so rounds to 0; 54 is half-way so rounds to 100. Turning the line vertically like a ladder makes it easy to judge if your number is half-way up (Figure 3.36).

Figure 3.36

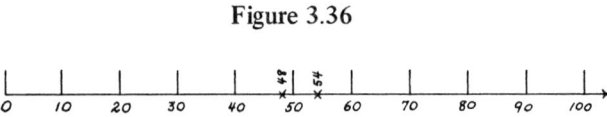

ABSOLUTE VALUE ON AN INTEGER NUMBER LINE (Figure 3.37)

Figure 3.37

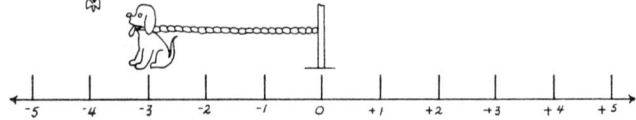

Stake a small cardboard dog on a 3-unit chain (call it a 3-yard chain) at 0. Discover that the farthest distance he can go along the line in either a positive or negative direction is 3 yards.

The distance is written as $|+3| = 3$ and $|-3| = 3$. It might be thought of as: "The distance the dog is away from 0 when he is at +3 is 3, and the distance the dog is away from 0 when he is at -3 is 3," but it is read mathematically as: "The absolute value of +3 equals 3, and the absolute value of -3 equals 3."

This activity may be extended to include a cat's wanderings: $|-3 + (+2)| = $? In this case, let each unit represent a city block. We see that the cat first went three blocks in a negative direction and then came back two blocks in a positive direction. By working the problem inside the absolute value signs, we determine that the cat stopped at -1, and is therefore 1 block in distance away from 0 (home). We now write this as: $|-3 + (+2)| = 1$. (Moving a cardboard cat along the line is an aid that gives added attraction to the lesson.) Another suggestion: Move a toy magnetic car along her route as mother goes shopping and does errands.

SLIDE RULE NUMBER LINES

Use two number lines of equivalent size that are similarly calibrated for adding and subtracting whole and fractional numbers. To add on the slide rule:

Understanding Number Symbols

Find the first addend on the A slide, and position the 0 on the B slide under it. Find the second addend on the B slide and read the sum above it on the A slide (Figure 3.38).

Figure 3.38

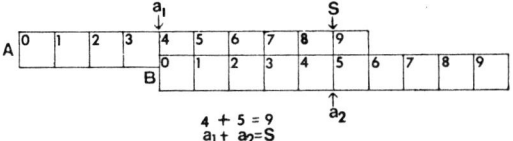

For a fraction slide rule, carefully mark off ½ and ¼ unit intervals on each of the slides and follow the directions given above for adding whole numbers (Figure 3.39).

Figure 3.39

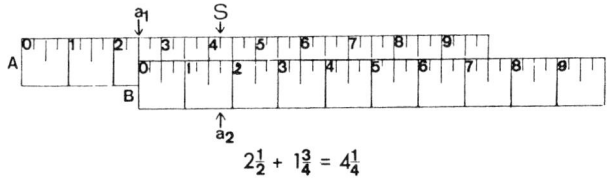

Subtraction is the inverse operation of addition. Find the minuend (sum) on the B scale. Match this with the subtrahend (known addend) on the A scale. Read the difference (missing addend) on B under the 0 on the A scale.

Integer Slide
Figure 3.40

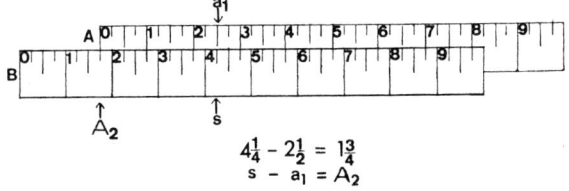

Figure 3.41

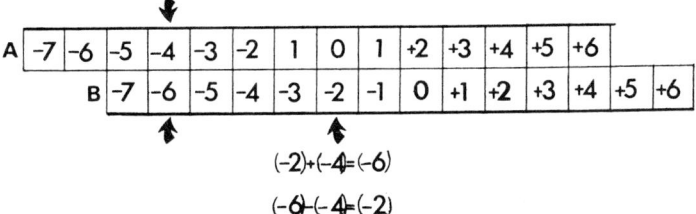

4.

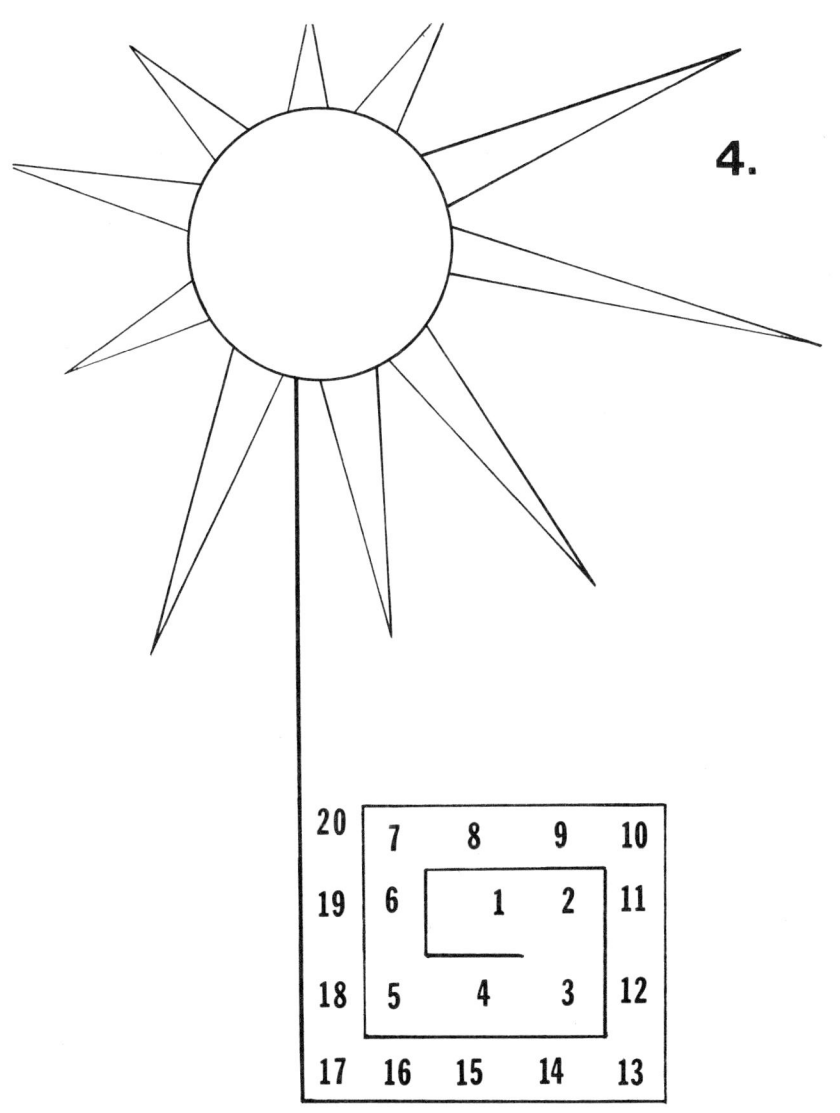

4

Bright Ideas For

The Counting Chart

—a welcome adjunct to every math room

Sometimes, one particular chart offers such a wide range and variety of relevant number situations that its value is immeasurable. Consider the Counting Chart, usually seen only in primary rooms, but which, with investigation, yields such a multitude of mathematical patterns and interrelationships that it should rightly be regarded as a powerful tool in any math room.

Counting is an order relationship. Number understanding is necessary for counting to be meaningful and not merely rote word-calling. Each number is one more than the preceding number. Early experiences with sets and number lines have helped develop order sense, and paved the way for using this compact form of ordered numerals, the Counting Chart.

All the fundamental binary operations are based on counting. The whole structure of addition is "counting one more." Multiplication's concern is how to count equivalent sets quickly. All the operations may be thought of as merely "fast counting." It is much faster to count by 10's, or by 5's, than by 1's. Although children are expected to learn the operational facts soon, so they do not have to resort to counting to solve problems, it is still a reliable fall-back method to a solution.

For pure discovery, *expose a class to a Counting Chart* often. Encourage the students to seek out patterns: What is the pattern of the rows? the columns? the falling diagonals (lines that fall as they proceed from left to right)? the rising diagonals? the multiples of each number? These explorations are exciting and lead to thoughtful generalizations. For example, all the rows on the chart have numerals with the same ending pattern as those in the first row. A child, realizing this, has acquired the whole secret of ordering all the numbers from 0 to 100 − and on to infinity.

When operational relationships have been personally discovered, the fundamental laws governing operations, the laws of order (commutative) and of grouping (associative) are better understood. Speed and accuracy increase.

This chapter tells and shows how to use the Counting Chart to:

emphasize the Base 10 system of counting

identify, analyze, and use patterns, families, and other relationships to reinforce or gain new ideas for addition and subtraction work

explore multiplication and division as organized multiple patterns

intrigue your pupils with column-choosing addition that produces Magic Sums and follows the numbers right off the chart

show counting charts can be different, and shows the use of part-charts

sift primes with Eratosthenes Sieve

For better understanding of an algorithm, return to the counting chart and study it carefully. Try to ascertain why a method works, and share discussions about the ideas and knowledge gleaned from the chart. It may well help to provide a simple explanation, as it did for students who made a discovery of a famous algorithm. (See Great Discovery, Fractions, Chapter 8.) Counting chart applications are found in many other places throughout the book. See Chapter 10, for a quick-switch of an ordinary Base 10 counting chart into a chart for many bases; Chapter 8, where it conforms explicitly to decimal fraction and percentage use; and Chapter 9, for Poison Square, Spill The Beans, and Other Games.

Figure 4.1

COUNTING CHART – A Compact Picture of Numbers From 1 to 100 (Figure 4.2)

Figure 4.2

1	2	3	4	5	6	7	8	9	10
11	12	13	14	15	16	17	18	19	20
21	22	23	24	25	26	27	28	29	30
31	32	33	34	35	36	37	38	39	40
41	42	43	44	45	46	47	48	49	50
51	52	53	54	55	56	57	58	59	60
61	62	63	64	65	66	67	68	69	70
71	72	73	74	75	76	77	78	79	80
81	82	83	84	85	86	87	88	89	90
91	92	93	94	95	96	97	98	99	100

VARIATION OF THE COUNTING CHART

Reasons for using:

(1) The digits, or 1-place numerals, 1 to 9, appear in a row alone: (2) The 3-place numeral, 100, also is alone in a row, and indicates the counting chart may be extended and that there is no end to our numeration system: (3) The tens (10, 20, 30 ... 90) are at the beginning of each decade – 20 with the twenties; 30 with the thirties, and so on; (4) adding to the tens is clearly seen – 12 is 10 and 2 more (Figure 4.3).

Figure 4.3

	1	2	3	4	5	6	7	8	9
10	11	12	13	14	15	16	17	18	19
20	21	22	23	24	25	26	27	28	29
30	31	32	33	34	35	36	37	38	39
40	41	42	43	44	45	46	47	48	49
50	51	52	53	54	55	56	57	58	59
60	61	62	63	64	65	66	67	68	69
70	71	72	73	74	75	76	77	78	79
80	81	82	83	84	85	86	87	88	89
90	91	92	93	94	95	96	97	98	99
100									

REMOVABLE NUMERAL COUNTING CHART (Figure 4.4)

A chart with removable numerals provides a different type of activity and more action on the part of the children, and it has a built-in testing facility.

A laminated, empty 100-square matrix is attached to cardboard or plywood of the same size by mystic tape around the edges. Small nails or hooks are tacked in each square. Small squares are punched and numbered 1 to 100. (Numbers have their own places in a counting chart – it is an ordered table. Children first practice them that way, but later the teacher puts one or two of the numerals out of place to see if the children can discover these and put them back in order.)

Figure 4.4

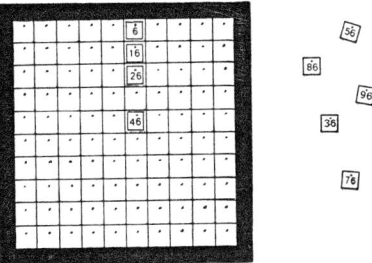

COUNTING CHART FROM 100 TO 200

As soon as the class is thoroughly familiar with the 1-100 chart, they can tackle the class project of filling in a 101-200 chart using a random method. This will test their knowledge of the pattern and of the ones and tens progression of the chart (Figure 4.5).

Figure 4.5

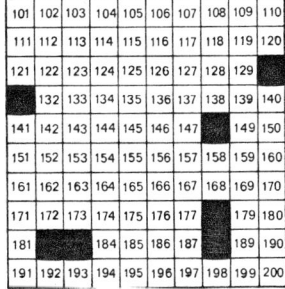

USING THE COUNTING CHART

1. Use the chart for counting procedures. When children are first learning to count to 100, utilize the intrinsic pattern of the chart to provide them with an unsurpassed help for exploring the simple, but efficient pattern of our decimal number system.

2. To give further refreshing practice to fix counting skills and competence in writing the numbers from 1 to 100 in order: (1) entirely cover the large counting chart; (2) have children write the first row (1 to 10); (3) expose this row on your chart and have them check their work; (4) have them write the next row; (5) expose for checking, and continue this procedure for all rows.

3. Have children count by 1's; by 2's; by 3's; by 5's; by 10's; by 2's

beginning at 7, and at 17, and discuss the pattern; count backwards by 1's, by 2's, by 5's, by 10's, or any set amount to suit the lesson or the capabilities of the children; count within certain limits, as, count by 4's from 16 to 32 (Figure 4.6).

Figure 4.6

1	2	3							
2	4	6							
8	11	14							
100	95	90							
16	20	24							

4. Use the chart for number activities that promote better awareness and recognition of numerals. Ask: Where is 47? Point to 74. Find 86. Show me the numeral that tells how many paws a kitten has; the numeral that tells your age. If zero is on your chart, ask a child to point out the numeral that lets us know how many kangaroos are in the room.

5. Use the chart for competence in reading and writing isolated numbers. Point to a numeral, ask a child to read it, write it, and compare it to the one shown on the chart. Try saying a number, having it found and written.

6. Use the chart to emphasize the importance of 10 in our numeration system. Lead the children to discover that this is a 10 by 10 chart. There are 10 in each row, and 10 in each column. After each 10 the ending digits repeat again. It is a Base 10 or decimal system. (*Decem* is Latin for ten.) Ask: what number means 2 tens and 6 ones? Point to the numeral that means 1 more than 10; one more than 20. A dozen is 12. How many tens and how many ones make a dozen? Show me that 20 is two sets of 10 (Figure 4.7).

Figure 4.7

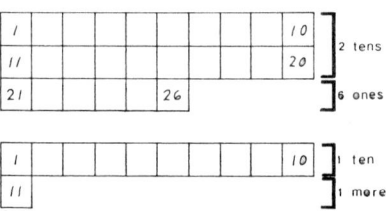

7. Have pupils learn the pattern of the chart. Each row numeral shows an increase of 1 in number value from its precedent. Each column numeral shows a 10 number value increase from the numeral above it. Count by 1's across the rows; by 10's down columns. Try counting by 100's. Fill in a new chart (Figure 4.8).

Bright Ideas For The Counting Chart

Figure 4.8

1	2	3	4				
10	20	30	40				
100	200	300	400				

8. Develop the special significance of the numerical structure of place value. Ten is the first 2-place number. You need two digits and two places to write it. Name the numbers that need only one digit and one place to write each. Name some numbers that need two digits and two places. There is one number written on the chart that needs three digits and three places to write it. It is the first 3-place number. What is it? Do you know the next 3-place number?

9. Develop a sense of number order. Have children look at the chart and read the teens in order. Ask: What number comes after 17? before 17? Which is the first number on the chart? the second? the 5th? the 17th? the last? Which is the greatest number shown on the chart? the least? Which is greater, 4 or 5; 9 or 19? 29 or 19? 47 or 74? Which is less, 6 or 7? 37 or 73? Show me your age, and your sister's age. Which number is greater? Who is older? Who is younger? Which number is less than the other? Show me the number of days in a week. There are 14 days until Thanksgiving Day. Show me 14 on the chart. Is there more than a week until Thanksgiving Day? More than two weeks. John has 2¢. He wants to buy a balloon for 5¢. Look at the chart and see if he has enough money.

10. Parts of the counting chart are used for oral or written work. Tell or write the missing numerals (Figure 4.9).

Figure 4.9

A.	1		3		6				10
B.	61							68	69
C.		22	23	24					

11. By covering certain numerals, associate "betweenness" with two numbers. Cover 23 and ask: What number comes between 22 and 24? Cover 29 and 30 and ask: What two numbers are between 28 and 31? (Figure 4.10).

Figure 4.10

1	2	3	4	■	6	7	■	9	10
11	12	13	14	15	16	17	18	19	20
21	22	■	24	25	26	27	28	■	■
31	32	■	34	35	36	37	38	39	40

71

12. Who are the neighbors? Say or write the missing numerals Laminated charts are easily erasable (Figure 4.11).

Figure 4.11

Before
35
65
85
14
95
30

Between	
16	18
29	31
48	50
63	65
97	99
55	57

After
29
49
17
37
40
11

13. Use the chart in like manner to provide practice with other essential math vocabulary that will be introduced throughout the Elementary and Junior High grades: above, below, begin, column, row, counting number, down, up, end, stop, start, greater, greater than, less, less than, least, right, left, digit, member, element, set, subset, next, numeral, number, order, over, under, pair, part, some, several, square, square number, square root, older, younger, equal, not equal, equivalent, not equivalent, odd number, even number, commutative law, associative law, unit. It is an excellent choice for the more difficult math terminology: fractions, decimals, prime number, composite number, percent and percentage, whole number, one half, one fourth, three fourths, one tenth, addend, sum, add, subtract, multiply, divide, factor, greatest common factor, multiple, least common multiple, divisor, least common divisor, product, quotient, dividend, other bases, Base 10, cardinal number, ordinal number, order, remainder.

14. To assist children in distinguishing the signs for greater than (>), and less than (<), have them think where the greatest amount of ice cream is in an ice cream cone. Turn the cone on its side to see the *greater than* sign and the *less than* sign (Figure 4.12)

Use >, or < to fill in the blank squares in Figure 4.13.

Figure 4.12

Figure 4.13

2	3
13	11
24	14
96	93

15. Have the pupils learn the reverse pattern of the chart. Each row number shows a decrease of 1 in value from the one on the immediate right. Each column number shows a decrease of 10 in numerical value

Bright Ideas For The Counting Chart

from the one below it. Count by ones from 10 to 1; Figure 4.15 by tens from 96 to 6 (Figure 4.15)

Figure 4.14

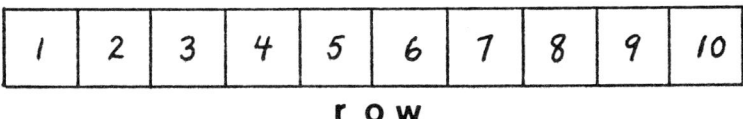

r o w

16. Adding one and ones. (1) Move the required amount of the first addend. (2) Continue moving along the row or rows the required amount of the second addend. The moves are always in the direction of greater numerical order (Figure 4.16).

Figure 4.15   **Figure 4.16**

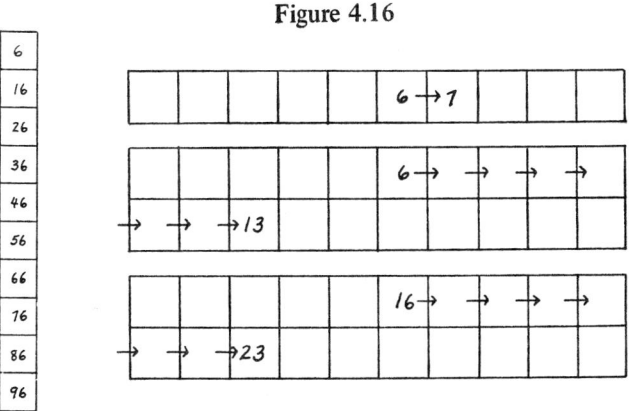

17. The order of the addends in Figure 4.16 have been changed to form the problems in Figure 4.17. The counting chart visually proves that the sums remain the same and that the commutative law holds for addition.

Figure 4.17

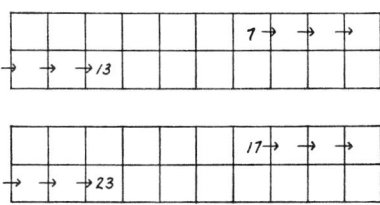

18. Adding or subtracting involving the "teens" are difficult for most children. Informal introduction to these facts, and extra practice, may be done on an individual counting chart constructed on two rows cut from the 100-square. Place sets of counters when adding, remove subsets when subtracting. Example: 17 - 9 = __. Counters are placed on squares from 1 to

17. The last nine are removed. Eight counters are left. Also, let "fingers do the walking" to 17, go back 9 and stop on 8 (Figure 4.18).

Figure 4.18

1	2	3	4	5	6	7	8	9	10
11	12	13	14	15	16	17	18	19	20

19. Cut and paste. Children cut apart, mix up, and then paste back together an individual part-chart as they arrange the numerals in counting order or add one more to all but the last number (Figure 4.19).

Figure 4.19

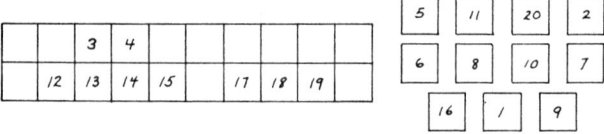

20. Subtracting ones: Start at the given sum, and using a counting procedure, move the required amount of the known addend along the row (or rows), in descending numerical order (Figure 4.20).

Figure 4.20

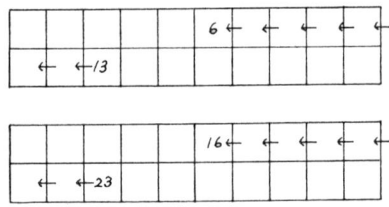

2 Adding 10 Coun or the chart to find the answer for 6 + 10. Discover tha the sum, 16 is directly below the first addend, 6. Count to find the answer for 16 + 10. Is this sum, 26, directly below 16? What is 10 more than 26? The answer is found just below 26. The pattern for adding 10 is "go down one square." Say: "Which is faster? Following the ten little arrows or following the one big arrow?" "What is the pattern for adding 10 on the Counting Chart?" Learn to add tens and then ones: 42 + 23. Think (42 + 20) + 3 = 62 + 3 = 65 (Figure 4.21).

Figure 4.21

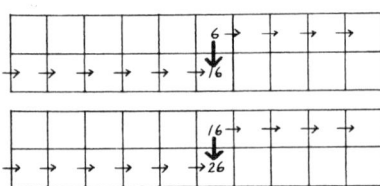

22. Adding 9: Count on the chart to find the answer for 6 + 9 = 15; 16 + 9 = 25; 26 + 9 = 35. Compare adding 9 with adding 10. It will soon become apparent to the children that to add 9 quickly, first add 10 and then go back 1 (Figure 4.22).

Figure 4.22

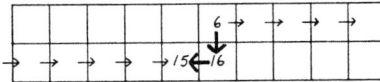

23. Add 20, then add 19. Do the same for 30 and 29, 40 and 39, et cetera. Make a rule for adding numbers ending in 9 (Figure 4.23).

Figure 4.23

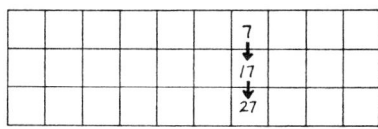

24. Subtracting 10. 16 - 10 =   Pupils should make a guess, then count back to prove. Try 26 - 10; 36 - 10; and so on. The action on the number line is the reverse of that of addition. The pattern is the inverse of the addition pattern. This articulation between the operations deserves attention (Figure 4.24).

Figure 4.24

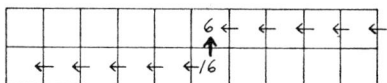

25. Subtracting 9. Notice how this relates to subtracting 10. 16 - 10 = 6. 16 - 9 = 7. 13 - 10 = *3*. 13 - 9 = *4* (Figure 4.25).

Figure 4.25

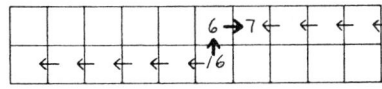

26. Find the addition patterns: 6 + 7 = 13. 16 + 7 = 23. 26 + 7 = 33. 16 + 27 = 43. 26 + 17 = 43.

27. Find the subtraction patterns: 13 - 6 = 7. 23 - 6 = 17. 33 - 6 = 27. 23 - 16 = 7. 33 - 16 = 17. 43 - 16 = 27.

28. Find the addition–subtraction interrelationships. (Family patterns): 6 + 7 = 13; 7 + 6 = 13; 13 - 6 = 7; 13 - 7 = 6.

29. Arrow arithmetic on the counting chart. Go one step in the direction indicated by each arrow. Try this without the chart. Examples: 7 → ↓ 18; 16 ↑ ↘ → 18; 25 → → → ↑ 18.

30. Estimation using the counting chart. Example: 29 + 42 = □. Think: 30 + 40 = 70, or think: 29 ↓ ↓ ↓ ↓ → → 71.

MAGIC SUMS AND THE INVERSE MAGIC OF SUBTRACTION:

More work with addition and subtraction on the Counting Chart gives the children a better insight into these operations and the intrinsic order of the chart itself. The experiences lead the children right off the chart to the numbers beyond, and convey to them the idea of a never ending system of numbers.

31. Choose a column on the chart. Choose two numbers from this column and add them. In which column can you find the sum? Try two different numbers from this same column. Which column contains the sum? (See Figure 4.26) If you commute the addends and add will you still find the sum in the same column?

32. Choose three addends from the 4 column, add them and determine the sum column. Try three different addends from this same column, and predict where the sum would be found. Example: 34 + 44 + 54 = 132. This sum is beyond 100, so it has led you off the chart. How can you tell in which column 132 would appear? (See Figure 4.27)

Figure 4.26

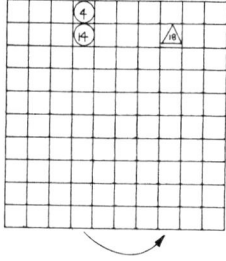

Figure 4.27

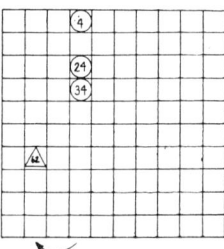

Figure 4.28

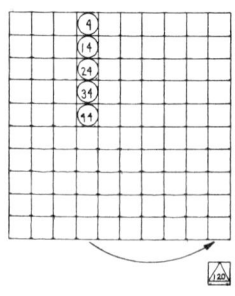

Bright Ideas For The Counting Chart 77

33. Could you predict where to look for the sum of four addends from the 4 column? Does this have any connection with the 4th multiplication table? How are 5 addends and the 5th table related? (Figure 4.28)

34. Choose a different column on the chart. Proceed as you did with the 4 column in #33.

35. Choose two columns, one addend from each, and see where the sum falls. Try again with different addends and predict where the sum will be found. Could you get a sum of 30 if you choose a number each from Columns 6 and 1? (Figure 4.29)

36. Choose three columns and an addend from each. Can you use multiplication to predict where the sum will fall? Why not?

37. Try subtraction. Choose two numbers from the same column and subtract. Did this difference appear in the 10 column? Why? Choose and try again.

38. Choose a different column, two numbers, and subtract. Can you say in your own words why the difference always comes in the 10 column? (Figure 4.29).

Figure 4.29

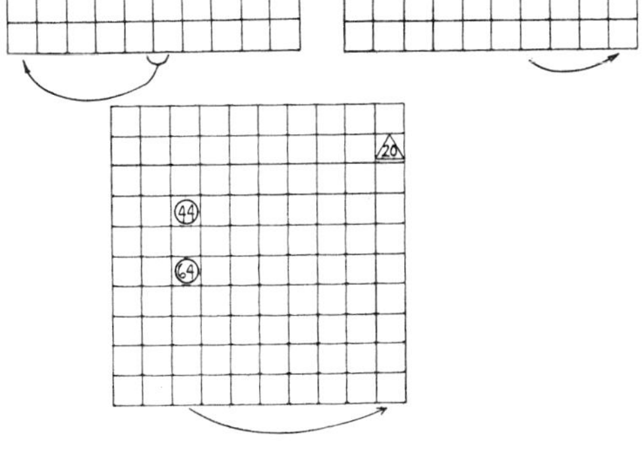

39. Choose two numbers from two different columns and subtract. Try several different pairs from the same two columns. Be careful! Example: Choose the 5 and 8 columns. Choose the numbers 85 and 38. 85 - 38 = 47.

Choose 65 and 48. 65 - 48 = 17. Choose 15 and 28. 28 - 15 = 13. Did the pattern change? Do you think the operation of subtraction is commutative?

40. Choose any two numbers in the 10 column. Add them. Where did this sum appear? Choose two other numbers and repeat. Now, choose two numbers from this column and subtract them. Where does the difference fall? Try again and again. Did you find that the 10 column has a different pattern from the other columns? Does zero have anything to do with the pattern?

41. Choose a number from the 10 column and one from any other column. Add them and see where the answer comes. Try a 10 column number plus a number from an entirely different column. Could you predict where this result would come? Does zero have anything to do with the answer patterns?

42. Investigate larger sums and numbers. Choose one addend from every column 1 through 10. If your sum were 535 and the counting chart was extended to include it, in which column would it appear? Could you always fit your sums into an extended counting chart? Tell in which columns you would find: 426? 439? How can you judge in which column 502, 631 or any particular number would fall? Is this similar to the way you judge whether a number is even or odd?

MULTIPLICATION AND DIVISION EXERCISES ON THE COUNTING CHART

43. Second table. Count by 2's and end up with the second table. Count by 3's for the third table; by 5's for the fifth table; by 7's for the seventh, etc. When counting by two's, a unit set of 2 squares for covering each count is helpful. For 6 x 2 = 12, the child must place the 2-square unit on the counting chart 6 times. Altogether his cover has touched 12 squares (Figure 4.30).

Figure 4.30

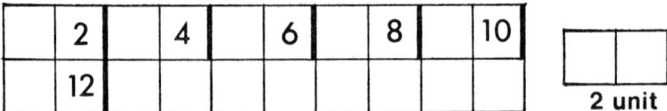

44. Do "whisper and aloud" counting for all tables. For the 3rd table: whisper 1, 2; say aloud 3; whisper 4, 5; say aloud 6; et cetera.

45. Tape three rows of the counting chart to the magnetic board. As the children play whisper and aloud, tap the chart on the whispered numerals, but place magnetic counters on the alouds. You now have a pattern of the multiples of three.

46. Play #45 on an empty grid to better see the pattern and the division by three.

47. Even numbers and odd numbers. Use the counting chart and count by two's for even numbers. Circle the numerals that represent the even numbers on the counting chart. Those not circled are odd numbers. Cross out (x) the odd numbers between 50 and 68. While looking at the counting chart count in unison: "Soft and Loud". Then do the same without looking at the chart (Figure 4.31).

Figure 4.31

1	2	3	4	5	6	7	8	9	10
11	12	13	14	15	16	17	18	19	20
21	22	23	24	25	26	27	28	29	30

48. If the chart includes a zero, the counting will be "Loud and Soft" (Figure 4.32).

Figure 4.32

0	1	2	3	4	5	6	7	8	9
10	11	12	13	14	15	16	17	18	19
20	21	22	23	24	25	26	27	28	29

49. Using the small 100-square paper, have pupils write the even numbers from 0 to 18, and observe the "ending digit pattern" (right digit). Have them describe the pattern, say the next ten even numbers, and tell whether or not any given number is a member of the set of even numbers (Figure 4.33).

Figure 4.33

0	2	4	6	8
10	12	14	16	18

50. Using small 100-square paper, have pupils write the odd numbers from 1 to 29. What discovery can be made about the right-hand digits of odd numbers? (Figure 4.34)

Figure 4.34

1	3	5	7	9
11	13	15	17	19
21	23	25	27	29

51. Color sets of even and odd squares. Cut the sets so they have a base of two. Notice that the *odd* sets do *not look even* across the tops (Figure 4.35).

Figure 4.35

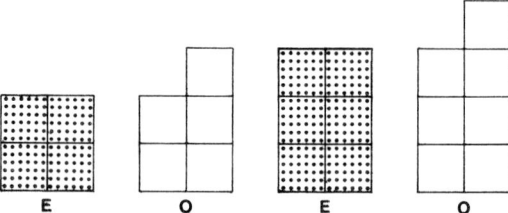

52. Adding evens and odds. To show that the sum of an even addend and an odd addend gives a sum that is odd, join a four-square and a five-square. Repeat with other sets. Join two odd sets and record whether the sum is even or odd. Repeat with the other odd sets. Try two sets of evens and record results. Return to the Counting Chart and choose addends from two even columns. Note whether the sum is an even or odd number. Do the same for even and odd number columns and for two odd number columns. Repeat the test many times. Test out subtraction in the same way. A remembering trick: adding any doubles will give you an even number, and the near doubles will always have odd answers. 7 + 7 = 14; 7 + 6 = 13; and 7 + 8 = 15. See Figure 4.36.

Figure 4.36

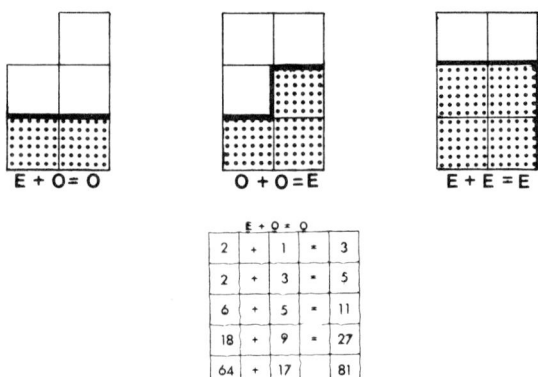

53. Test out and record similar charts for multiplication of odd and even numbers.

54. Counting by 5's. Count the numerals that have been covered in order, 5, 10, 15, 20, et cetera. Count "Soft and Loud" — the multiples of 5 are the loud ones. Write the missing numerals in individual charts. Color the multiples of 5. Write the multiples of 5 from 5 to 50 in a strip of ten squares. Then arrange them in order and cut and paste them on another strip of blank squares. Cover the multiples of 7 on the counting chart. Counting the 7's "Soft and Loud" helps children learn the 7th table. Count up by 7's and count down by 7's. This exercise is adaptable for any multiple pattern or multiplication table (Figure 4.37).

Figure 4.37

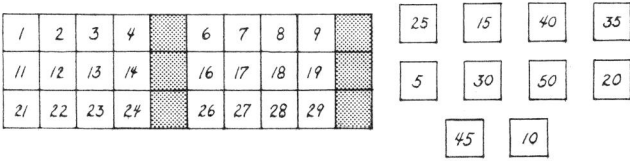

55. Even division on the chart. 6 ÷ 2 = ? Start at the product and count back as many equivalent sets of 2 as possible to land exactly on 1. Using a 2-square cover unit as he counts, the child sees that he must use this cover 3 times to count 6. Thus, 6 ÷ 2 = 3. This is a measurement type of division. 6 is measured by a 2-unit and this unit was placed 3 times to measure 6 completely. Comparing multiplication and division shows the interrelationship: each time the child was dealing with equivalent sets of 2; and whenever working with either multiplication or division, the numbers 2, 3, and 6 are associates. The same would be true of 7, 8, and 56; 7, 9, and 63; 4, 8, and 32; et cetera. Point these triads out to the children as they learn their multiplication facts, and division facts will be easier to learn (Figure 4.38).

Figure 4.38

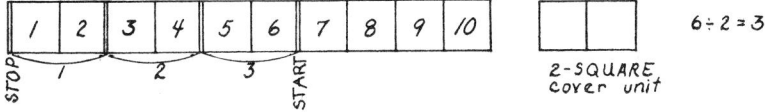

56. Uneven division on the chart. Since division means the sets must be equivalent, whenever there are not enough remaining to measure another equivalent set, there is a remainder. 7 ÷ 2 = 3, with a remainder of 1. After the 2-unit set has been placed 3 times, the 1 square remains unmeasured since 1 is not enough to make a set equivalent to the 2-unit set. It is enough to make one half of the unit set; it can be measured as ½, or it may be left as a remainder of 1 (Figure 4.39).

Figure 4.39

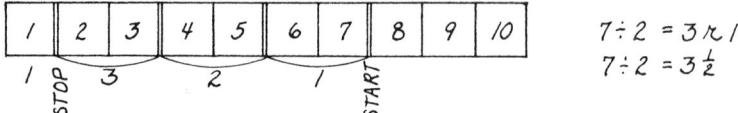

57. Multiplication-Division table of 9's. Circle all numbers on the chart whose digits add up to 9. End up with the ninth table. Notice where the multiples of 9 appear on the chart. Draw diagonal lines through them. Nine divides 10 with *1* remainder. Does this account for the move of *1* to the left of each succeeding row? Should the eleventh table move *1* to the right? Test out your guess (Figure 4.40)

Figure 4.40

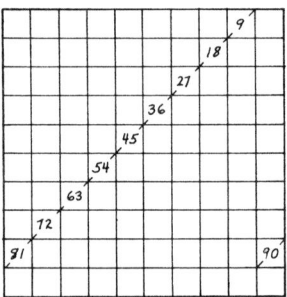

58. Third table. Circle numbers divisible by 3. End up with the third table and other multiples of three. Draw diagonals. Find the differences between the numbers along the diagonals. How do you explain a difference of 9? of 12? Add the digits of each multiple. If you get a sum with two digits add these until you have a one-digit sum. Notice the pattern formed by the ones-place-digits in each multiple along the rising diagonal; the tens-place-digits.

> 3 divides 10 with *1* remainder. Each row has a shift of 1 to the left. What other table has the same pattern? Can you easily find the 9th table within the 3rd table? Why? What other table is hidden here? (Figure 4.41)

Figure 4.41

59. Sixth table. Circle all even numbers (2nd table) that are multiples of 3. Draw diagonals. Notice the shift of 4 to the left in succeeding rows.

Find the differences between numbers along the diagonals.

Add the digits of each multiple of 6. How do these sum patterns compare with the sum pattern of the multiples of 3?

Variation: Tape counting chart on magnetic board. Place black counters on multiples of 2, and red ones on multiples of 3. What happens on multiples of 6?

Both 2 and 3 divide 6 with no remainder so they are both factors and divisors of 6 (Figure 4.42).

Figure 4.42

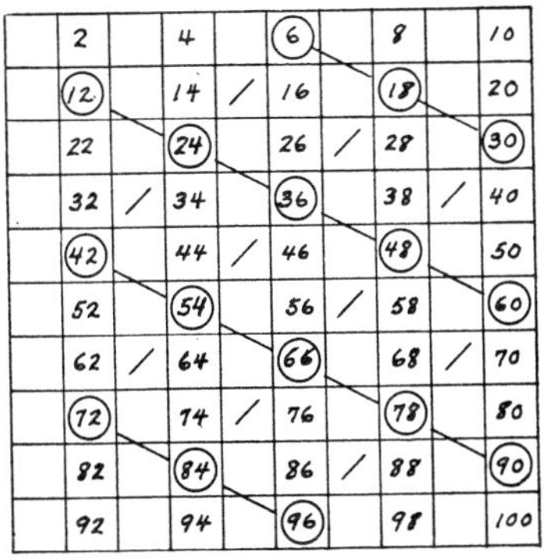

2 4 6 8 10
12 14 16 18 20
22 24 26 28 30
32 34 36 38 40
42 44 46 48 50
52 54 56 58 60
62 64 66 68 70
72 74 76 78 80
82 84 86 88 90
92 94 96 98 100

60. Consider multiples of 2, 3, and 6, and prime numbers. (1) Write the primes on an empty 100-square grid. (2) Mark a cross (+) where the multiples of 2 would be located on the grid. Where are all the primes in relation to the multiples of 2? Why? (3) Mark a small x where the multiples of 3 would lie on the grid. Where are all the primes in relation to some of the multiples of 3? Why? (4) The multiples of 6 are now marked with a star (*). Where are all the primes (except 2 and 3) in relation to some multiples of 6? Must every prime have a multiple of 2, of 3, and of 6 for a neighbor? Test to 200 (Figures 4.43, 4.44).

Figure 4.43

	2	3		5	*	7			
11	*	13				17	*	19	
		23	*					29	*
31					*	37			
41	*	43				47	*		
		53	*					59	*
61					*	67			
71	*	73					*	79	
		83	*					89	*
					*	97			

Figure 4.44

101	*	103				107	*	109	
		113	*						*
					*	127			
131	*					137	*	139	
			*					149	*
151					*	157			
	*	163				167	*		
		173	*					179	*
181					*				
191	*	193				197	*	199	

61. Fifteenth table. Combine 3rd and 5th tables to find the common multiples of 3 and 5.

This table is very useful when dividing by teen numbers. Estimate that $102 \div 14 = 7^+$ because $105 \div 15 = 7$. It is easy to learn when you think of a clock (Figure 4.45).

Figure 4.45

				5				10	
				15				20	
				25				30	
				35				40	
				45				50	
				55				60	
				65				70	
				75				80	
				85				90	
				95				100	

62. Vertical patterns. Do 2, 5 and 10 all divide 10 without a remainder? Is there a diagonal shift in the patterns of these tables?

63. Fourth table. Circle all multiples of 4. Combine the 3rd and 4th tables and circle the common multiples. What table results?

64. Square numbers. Locate the square numbers. Does any pattern develop? How many steps are there between 1 and 4? 4 and 9? 9 and 16? Continue. What did you discover? Predict the next square number.

Figure 4.46

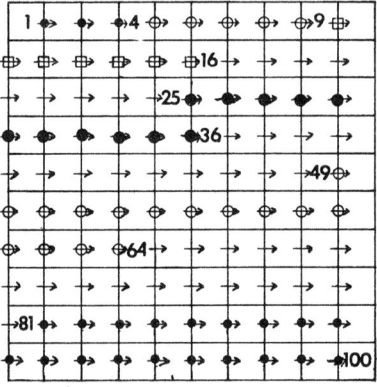

See Chapter 1, A Study in Design. Figures 1.25 and 1.26 show the sum of the odd numbers form the square numbers. Is the pattern on the counting chart basically the same? $1 + 3 = 4; 4 + 5 = 9; 9 + 7 = 16; 81 + 19 = 100$ (Figure 4.46).

Bright Ideas For The Counting Chart 85

65. Counting Chart Patterns. Name the pattern shown on each grid (Figure 4.47).

Figure 4.47

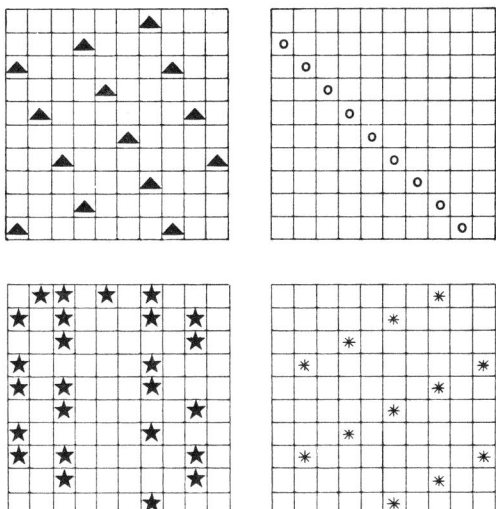

66. Eratosthenes' Sieve for Finding Primes. To find all prime numbers from 1 to 100, first write the numerals 1 to 100. 1 is the identity factor and not a prime number. 2 is the first prime number. Sift the multiples of 2 out of your chart by scratching out every multiple of 2 except 2. 3 is the next prime number. Scratch every multiple of 3 except 3. 5 is the next prime. Scratch every multiple of 5 except 5. 7 is the next prime. Scratch all multiples of 7 except 7. All those that remain not sifted out are prime numbers. How many primes are there from 1 to 100? Why was it not necessary to test above 7 when finding the prime numbers from 1 to 100? (It is not necessary to test above the square root of any number to find all the factors of that number, since the factors below the square root pair up with the factors above the square root. When testing for primes, test with only the prime numbers below the square root. 10 is the square root of 100 and the primes below 10 used for testing were 2, 3, 5, and 7.) Use the overhead projector and colored overlays to sift out multiples of 2, 3, 5, and 7, leaving the 25 primes. These overlays may also be used in pairs to find common multiples (Figure 4.48).

Figure 4.48

1	(2)	(3)	4	(5)	6	(7)	8	9	10
(11)	12	(13)	14	15	16	(17)	18	(19)	20
21	22	(23)	24	25	26	27	28	(29)	30
(31)	32	33	34	35	36	(37)	38	39	40
(41)	42	(43)	44	45	46	(47)	48	49	50
51	52	(53)	54	55	56	57	58	(59)	60
(61)	62	63	64	65	66	(67)	68	69	70
(71)	72	(73)	74	75	76	77	78	(79)	80
81	82	(83)	84	85	86	87	88	(89)	90
91	92	93	94	95	96	(97)	98	99	100

COUNTING CHARTS CAN BE DIFFERENT —
Count Down and Subtract

These charts help children who tend to transpose digits in a numeral. Because of its position on the chart, some children read 47 as 74. Suggestions: Children (1) start at 1 and stay within the green line as they count to 100, (2) start at 100 and count backward as they wind themselves up; (3) write the prime numbers in red, and play a game using flashcards, problems, dice or spinner, to move counters along the squares to 100. Landing on a prime gives a chance to move along a diagonal line as far as it goes, but watch out — an opponent can send a player back to 1 if he lands exactly on his square; (4) play Lost Numeral or Lost Prime (a real challenge if the spiral path is not drawn in and children must discover the pattern themselves); (5) construct their own Lost Numeral, Extra Numeral, or similar games for classmates to play (Figure 4.49).

Figure 4.49

100	99	98	97	96	95	94	93	92	91
90	89	88	87	86	85	84	83	82	81
80	79	78	77	76	75	74	73	72	71
70	69	68	67	66	65	64	63	62	61
60	59	58	57	56	55	54	53	52	51
50	49	48	47	46	45	44	43	42	41
40	39	38	37	36	35	34	33	32	31
30	29	28	27	26	25	24	23	22	21
20	19	18	17	16	15	14	13	12	11
10	9	8	7	6	5	4	3	2	1

Bright Ideas For The Counting Chart 87

5.

10 *is the key* to REGROUPING

5

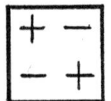 # Solve Your Addition and Subtraction Problems

−stress place value − there are pitfalls in regrouping

Addition and its counterpart, subtraction, are built on the foundations of number concept and place value. Yet, even when the beginner's program has been thorough, and no complications occurred when the binary operations were introduced, parents, children, and teachers all recognize the subsequent pitfalls of learning facts and regrouping tens.

To the dismay of upper grade teachers, many an older child resorts to counting on his fingers instead of quickly recalling needed addition facts. Often when this child must subtract, even his fingers fail him, either because he does not understand their role in the subtraction process, or because he has never learned to count down correctly.

During the Middle Ages, when Roman numerals were in common use in Europe, counting was the accepted method for adding or subtracting. Beads, on a counting frame or abacus, were moved one at a time in a direction to show "more" beads on a particular wire, or moved in the opposite direction to present "less" beads on the wire. A child could do, see, and understand the operation − one on − adding 1, − one off − subtracting 1.

The gradual acceptance of the Arabic numerals, with their zero symbol and place value scheme, freed the math worker from dependence on the counting frame and gave him the chance to use the more efficient modern-day algorithm for performing addition. The whole structure of the operations remained the same — adding is "counting one more," subtraction is "counting one less," but the method of performing these operations changed from a concrete working with beads to an abstract working with numerals.

Realizing this, teachers of primary children provide concrete manipulative materials at the initial stages. Adding one and subtracting one are presented as joining one object to a set to give it one more member, or removing one from a set to leave it with one fewer member.

To bridge the gap between concrete material and abstract number, number lines and counting rows are used. Particularly effective is the row of footprints that go up a hill. Addition and subtraction problems are presented at the same time, the signs indicating whether to walk up one or down one. The sign (+) plus means "count up". The sign (-) minus means "count down." This simple, pre-number line leads to use of the regular, more abstract number line, and then on to daily facts practice without the support of the number line.

Highlights of this chapter include:

simple flashcards children construct for their individual needs

using a slide to help understand the construction and purpose of the addition-subtraction chart, and the related equations of these operations

using squared paper and long zigs and short zags to turn out quick practice material

folding-strips for addition and subtraction practice

Many other suggestions are offered in this chapter, in Chapter 4, where counting chart addition activities are explained, and also in Chapter 9, where pertinent, teachable games are listed, with hopes that through their use, children will have better command of their facts and will better understand that: subtraction is the inverse of addition; adding one is the inverse of subtracting one; if $4 + 1 = 5$, then $5 - 1 = 4$; and, if, when adding, we can build a set of 10 ones, wrap it up and rename it as 1 ten, then we can also unwrap 1 ten and rename it as a set of 10 ones when subtracting.

Figure 5.1

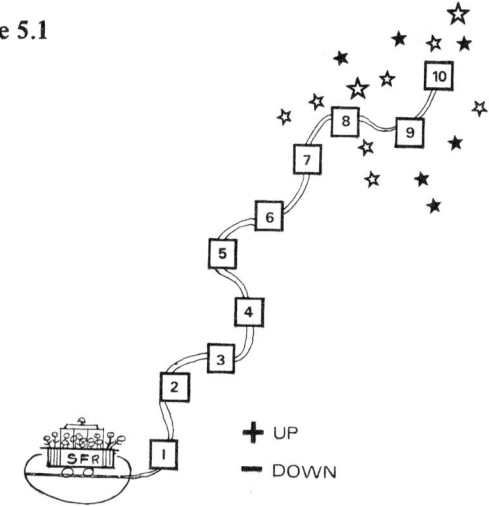

+ UP
− DOWN

DOMINOES

Suggestions:

1. Children draw dots on the halves left blank to make each domino have a total of 10 dots.

2. Children draw dots on both blank halves so each domino will show a different combination with the sum of 7. (Or any given sum.)

3. Children play dominoes using large dotted squares. (To prepare, paste dotted squares on cardboard and laminate.)

4. Children learn to estimate by using the domino doubles. 5 + 5 = 10. Is 5 + 2 greater than or less than 10? Is 5 + 6 greater than or less than 5 + 5? (Figure 5.2)

Figure 5.2

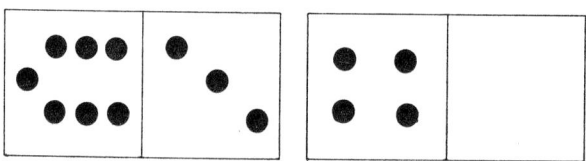

"OPEN - UP" FLASHCARDS

Here's how:

Use a strip of four large squares. Fold the top square and the bottom square backward along the lines. Children draw dots and number the cards as shown in Figure 5.3. They make only the facts they need to learn.

Solve Your Addition and Subtraction Problems

Figure 5.3

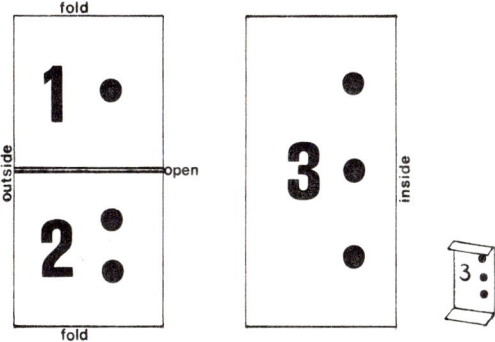

SUBSETS TEACH FAMILIES OF FACTS

Draw the sets and have children write 4 problems for each (Figure 5.4).

Figure 5.4

2 + 4 = 6					
4 + 2 = 6					
6 - 2 = 4					
6 - 4 = 2					

SUBTRACTING FROM 5

A 5 x 5 array is cut to show all the facts with 5 as a sum. Use similar arrays for other sums. Also, cut squares apart, show the children 4 squares and ask: "Can you take 5 from this set? Can you work this problem 4 - 5 = ?" (Figure 5.5)

```
5 - 0 =     0 + □ = 5
5 - 1 =     1 + □ = 5
5 - 2 =     2 + □ = 5
5 - 3 =     3 + □ = 5
5 - 4 =     4 + □ = 5
```

Figure 5.5

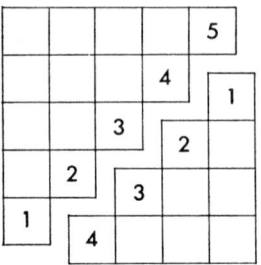

PICTURE ARRAYS

Find the unknown sums for picture arrays. Write 4 problems for each picture. 6 + 3 = □; 3 + 6 = □; □ - 3 = 6; □ - 6 = 3. (Figure 5.6)

Figure 5.6

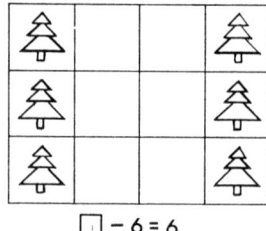

INVERSE OPERATIONS

Addition and subtraction facts are taught as inverses. The child starts at 7, goes to 13, and back down to 7. Each child may write a different stair problem for a bulletin board border (Figure 5.7).

Figure 5.7

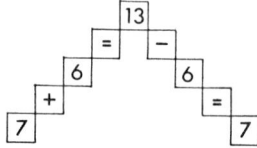

Encourage children to suggest and cut squares to arrange other balancing problems for a bulletin board display (Figures 5.8, 5.9).

Figure 5.8

Figure 5.9

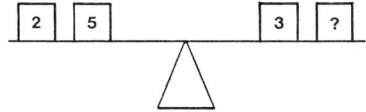

SHOW-ME-STRIPS HELP SOLVE NUMBER STORIES

Each child is supplied with a strip of large squares and counters, (lima beans, buttons, checkers, bingo counters, etc.). As the teacher tells a simple story, the children show that they understand the numbers involved by placing counters on their strips. These daily stories are supplementary exercises for practicing both arithmetic and reading skills.

When working with number stories (story problems) at any grade level, it is good policy to begin with very small numbers and then repeat the same stories over again with larger numbers. To make the stories more enjoyable and personalized, select a favorite theme and let the children help create the stories. The samples given are all about Banjo, the colorful little rooster and his rooster friends, Brewster and Lazybird. As the teacher tells the story and says, "Show me," the children are to respond with the correct number of counters on their strips. Numbers and equations are written on the chalkboard and answers given immediately. Later, the same number stories are written in simple language to be reused as a reading lesson, or to provide an arithmetic review, and to check and evaluate their achievement.

1. Banjo, Brewster, and Lazybird were roosters. They lived on Farmer Jack's farm. How many roosters lived on the farm? (Show me.)

2. One morning, Banjo woke up early. He jumped up on a stump and crowed 3 times. Brewster woke up and crowed 2 times. How many times did the roosters crow altogether? (Show me. First put on enough counters to show the number of times Banjo crowed. Then put on more to show how many times Brewster crowed. Does this show how many times both crowed altogether?)

3. Banjo liked to tip-toe through the tulip bed. He tried to be careful, but he did step on 1 big yellow tulip and 4 red tulips. How many tulips did he smash? (Show me.)

4. Lazybird went looking for corn to eat. He scratched in the radish bed and dug up 2 red radishes. Then he scratched in the onion patch and dug up 2 white onions. What a crazy bird was Lazybird. How many vegetables did he dig up? (Show me.) See Figures 5.10, 5.11.

Figure 5.10

Figure 5.11

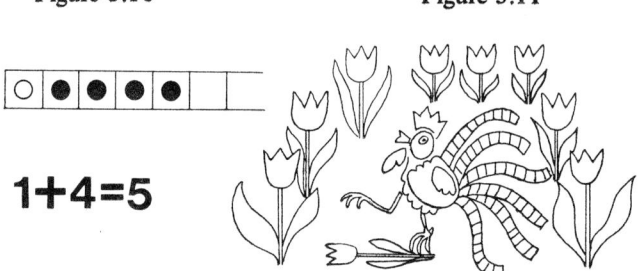

1+4=5

ADDITION-SUBTRACTION CHART (Figure 5.12)

Use the 100-square grid. First addends are written down left side. (Use blue felt marking pen.) Second addends are written across top. (Use red felt marking pen.) The sums may be written with a black felt marker. Pupils may construct addition charts on the small-size 100-square for their own use at the same time the big chart is being prepared for class use. (See suggestions which follow.)

Figure 5.12

+	0	1	2	3	4	5	6	7	8	9
0	0	1	2	3	4	5	6	7	8	9
1	1	2	3	4	5	6	7	8	9	10
2	2	3	4	5	6	7	8	9	10	11
3	3	4	5	6	7	8	9	10	11	12
4	4	5	6	7	8	9	10	11	12	13
5	5	6	7	8	9	10	11	12	13	14
6	6	7	8	9	10	11	12	13	14	15
7	7	8	9	10	11	12	13	14	15	16
8	8	9	10	11	12	13	14	15	16	17
9	9	10	11	12	13	14	15	16	17	18

CONSTRUCTING INDIVIDUAL ADDITION CHARTS

Each child will need a small 100-square grid and a right-angle slide. A simple addition equation is suggested, for example, 3 + 4 = 7. The slide is placed

Solve Your Addition and Subtraction Problems 97

around the first addend, 3, so it has a "floor" under it and a "wall" behind it. The slide is then pushed along the line until the "wall" is behind the second addend, 4. The sum, 7, is written on the chart in the corner square. This procedure is repeated until all the chart is filled.

This right-angle slide is used to show addends and sum for any addition fact. Later, use fingers to run from each addend to intersect at the sum (Figure 5.13).

Figure 5.13

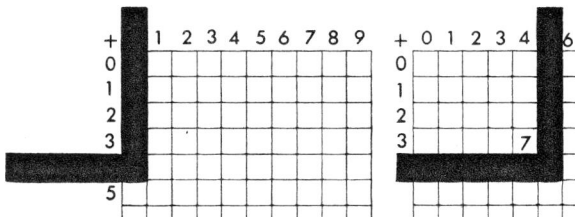

Correlating Addition and Subtraction

Use the right-angle slide for subtraction. Find the sum in the correct addend row, and locate the missing addend at the top of the chart. 8 - 3 = 5. (Figure 5.14)

Figure 5.14

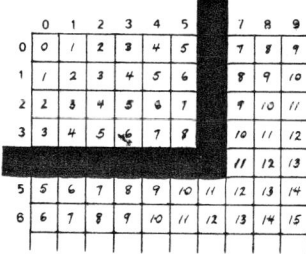

ADDITION AND SUBTRACTION PRACTICE STRIPS

Individual fact strips are made by the pupils. When folded along different lines the strips illustrate all the basic number combinations with the sum of 10. Dot strips and abstract numeral strips are used in the same way as the picture strips. When the strip is folded in half it illustrates the doubles fact 5 + 5 = 10. Children should think that 5 + 4 < 10, and that 5 + 6 > 10 (Figure 5.15).

Figure 5.15

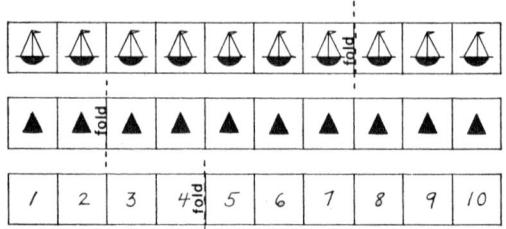

The strips may be fashioned to add to any sum (Figure 5.16).

Figure 5.16

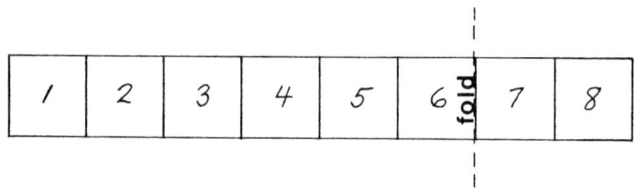

For individual instruction cut and use only a part of the addition table for practicing needed facts. Fill in sums, addends, or both (Figure 5.17).

Figure 5.17

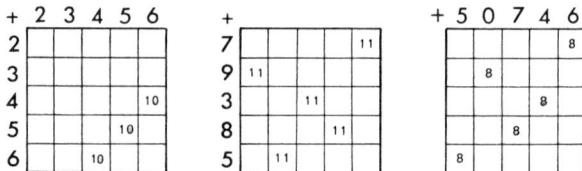

ADDING TO SPECIFIC SUMS

Facts that add to a specific sum are found along a rising diagonal. Other patterns include: the doubles are found along the falling diagonal, and the chart is commutative about this diagonal; zero is the identity element; subtraction is the inverse of addition (Figure 5.18).

Solve Your Addition and Subtraction Problems

Figure 5.18

+	0	1	2	3	4	5	6	7	8	9	10
0	0	1	2	3	4	5	6	7	8	9	10
1	1	2	3	4	5	6	7	8	9	10	
2	2	3	4	5	6	7	8	9	10		
3	3	4	5	6	7	8	9	10			
4	4	5	6	7	8	9	10				
5	5	6	7	8	9	10					
6	6	7	8	9	10						
7	7	8	9	10							
8	8	9	10								
9	9	10									
10	10										

PICTURES CHECK THE FACTS

1. Use three squares for each fact. Draw a picture on the back of each card. Cut off the answers. The child matches problems and answers, and checks the work by turning the cards over to see if the pictures are matched up. A robin's legs would be wrong on a boy. Use for more difficult problems in any operation (Figure 5.19A).

2. Cut two sets of 9 squares, one to fit over the other. On the bottom set write addition problems, and on the top set the corresponding answers. Also, draw a simple picture on the top set and cut the square apart. As the child matches problems and answers, the picture appears (Figure 5.19B).

Figure 5.19A   **Figure 5.19B**

TENS CHART

A tens chart is cut from the large 100-square matrix. Magnets are attached to the back of the chart, and to the individual strips for use on the magnetic board. Pupils hang the strips to make sums of 10. Both addition and subtraction practice may be performed.

The commutative law may be illustrated by hanging the 6-strip under the 4, and the 4-strip under the 6.

The associative law may be illustrated by hanging the 7-strip and the 1-strip under the 2 to sum to 10, and then hanging the 1-strip and the 2-strip under the 7. If $(2 + 7) + 1 = 10$, then $2 + (7 + 1) = 10$.

Show all the combinations with sums of 5; do the same for sums of 7, or 8, etc.

Place the entire chart together. Subtract from 10, by removing a 6-strip. Record. Remove other strips and record.

10 - 6 =	10 - 5 =	10 - 1 =
10 - 4 =	10 - 9 =	10 - 10 =
10 - 7 =	10 - 3 =	10 - 2 =

Use the same procedure for subtracting from 5, 7, 8, etc.

Place the 7 strip below the 6 strip. The child sees that: $6 + 7$ is 3 beyond 10; $6 + 7 = 13$; $7 + 6 = 13$; 13 is 10 and 3 more (Figure 5.20).

Figure 5.20

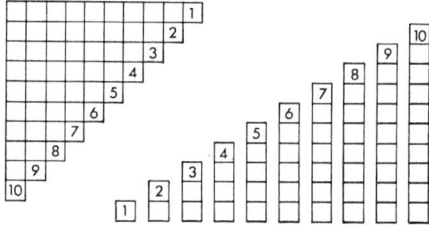

STARS AND ZIGZAGS

Write an addition fact for each row (Figure 5.21).

Figure 5.21

LONG ZIGS AND SHORT ZAGS

Write a subtraction equation for each row (Figure 5.22).

Figure 5.22

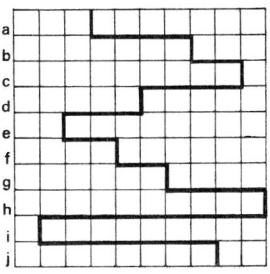

TRIANGLES AND A PAIR OF THOSE CROOKS

Write and solve an addition equation for each row. Also, write column equations. Examples: Row 1: 3 + 4 + 3 = 10; Column 1: 9 + 1 = 10. See Figure 5.23.

Figure 5.23

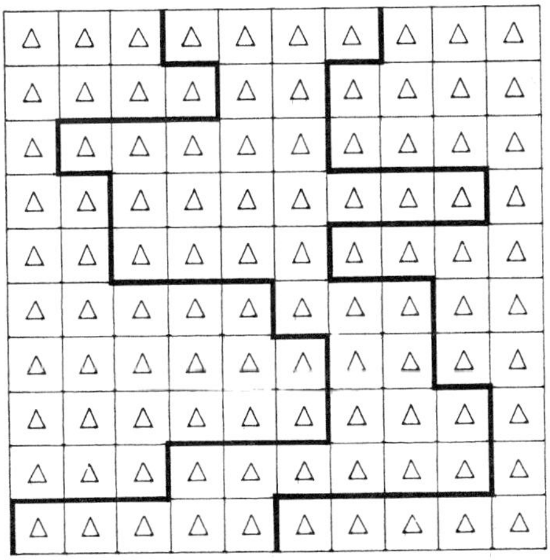

ARE YOU A COMMUTATIVE ADDER?

Sets of squares are numbered, colored and cut out.

1. Place the squares in stairstep order.
2. Count down the stairs and up again.
3. Place finger on each 2. Say 2 + 0 = 2; 2 + 1 = 3; 2 + 2 = 4, etc. Do the same for 3, etc.
4. Subtract 10 - 1 = 9; 9 - 1 = 8, etc.

5. Place sets end-to-end to illustrate addition.
6. Place sets together and compare for subtraction. 10 - 9 = 1, 10 - 8 = 2; etc. See Figure 5.24.

Figure 5.24

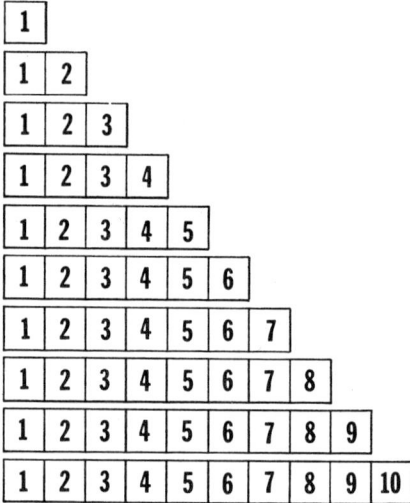

Show the commutative property by placing two sets beside the ten set, and then changing the order. 4 + 3 and 3 + 4 both equal 7. See Figure 5.25.

Figure 5.25

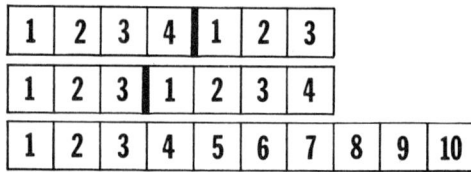

ASSOCIATIVE PROPERTY FOR ADDITION

Three sets of colorful flowers, birds or other pictures, are pasted on a 10-row. First, hands are used to enclose the first two numbers to be added. This sum plus the third addend completes the work. The hands are then used to show a different grouping arrangement which by the associative property produces the same sum. Parentheses take the place of hands (Figure 5.26).

$$(3 + 1) + 2 = 3 + (1 + 2)$$
$$4 + 2 = 3 + 3$$
$$6 = 6$$

Figure 5.26

The Associative Law may be illustrated by coloring sets of squares (Figures 5.27, 5.28).

Figure 5.27

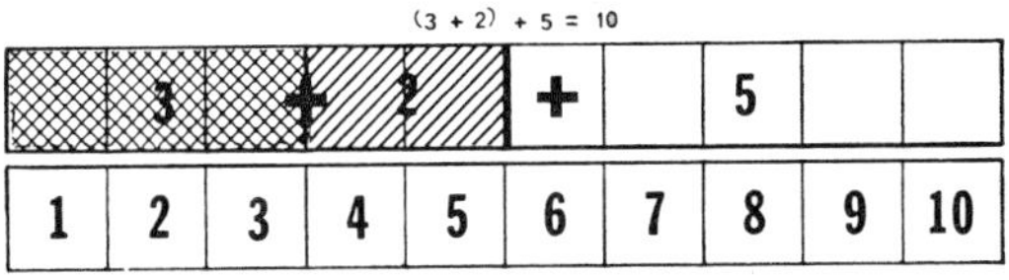

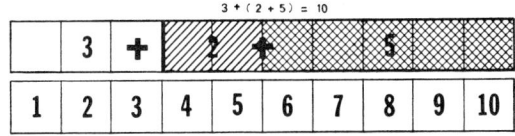

Figure 5.28

CALENDAR

The calendar is a type of counting chart. How is it different from the regular Base 10 counting chart? The row increase is 1; the column increase is 7 instead of 10. Where are the answers to 1 + 7? 8 + 7? 9 - 7? 19 - 7? Where are the multiples of 7? What is (4 + 7) + 1? Try adding 6 and 8 to the calendar numbers. Make a new calendar each month. Let the children write their names on tags to tack on their birth dates (Figure 5.29).

Figure 5.29

Sun.	Mon.	Tue.	Wed.	Thu.	Fri.	Sat.
				1	2	3
4	5	6	7	8	9	10
11	12	13	14	15	16	17
18	19	20	21	22	23	24
25	26	27	28	29	30	31

Addition Practice on the Counting Chart and Calendar.

Cut a 4 by 4 array from the large 100-square. Cut out the 4 center squares to leave a frame. Place the frame on either the counting chart or calendar to show a 2 by 2 array. Add across diagonals. Example on the counting chart:

6 + 17 = 7 + 16
23 = 23

Example on the calendar:

6 + 14 = 7 + 13
20 = 20

Multiply crisscross and note results.

Cut a 5 by 5 frame, to show a 3 by 3 array. (Figure 5.30)

Example:

6 + 14 + 22 = 8 + 14 + 20
42 = 42

Add the middle column and the middle row. Are these sums always equal? Why?

Add each of the columns. What are the differences between these sums? Explain.

Add each of the rows. What are the differences between these sums? On the calendar why is the difference always 21 between consecutive rows? See Figure 5.30.

Figure 5.30

↘				↙
	6	7	8	
	13	14	15	
	20	21	22	
↗				↖

TEN IS THE KEY – PLACE VALUE AND REGROUPING

Three colored sets, representing three addends, are joined to illustrate addition with regrouping. (5 + 4) + 3 = (5 + 4 + 1) + 2 = 10 + 2 = 12. Place the colored sets on a set of 20 blank squares to show the importance of making a ten (Figure 5.31).

Figure 5.31

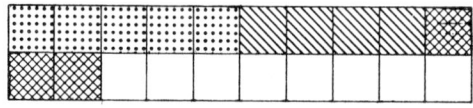

Solve Your Addition and Subtraction Problems

Related facts are shown with sets of colored squares (Figure 5.32). Review these often.

$3 + 5 = 8$ $5 + 3 = 8$
$8 - 3 = 5$ $8 - 5 = 3$

Figure 5.32

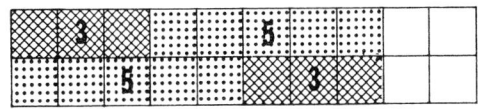

FLASHCARDS THAT CHILDREN CONSTRUCT FOR THEIR OWN NEEDS (Figure 5.33)

Figure 5.33

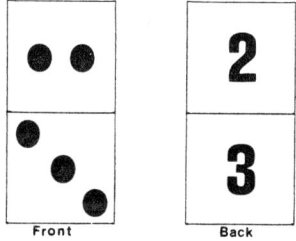

PLACE VALUE AND ADDITION

Quickie: Do you have to add or subtract to fill in blanks across each row? See Figure 5.34

Figure 5.34

a_1	a_2	a_3	Total
300	40	2	
400		9	469
800	70	1	
900			986

First add ones, then tens, then hundreds (Figure 5.35).

Figure 5.35

+	H	T	O
103 =	1	0	3
15 =		1	5
340 =	3	4	0
=			

TENS AND ONES

Subtraction with regrouping may be introduced with 10-strips and ones, used to work these problems (Figure 5.36).

1.	2.	3.	4.
24	23	22	21
- 2	- 2	- 2	- 2

For each problem, the correct number of tens and ones are shown and the child must remove 2 ones. What must happen in Problem 4?

Record as:

```
1 ten      11 ones       1    11
2 tens     1 one         2    1
                   
           2 ones    -       2
1 ten and  9 ones       1    9
```

1 ten = 10 ones.
1 from the next higher place gives you 10 of what you need.

Figure 5.36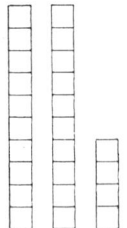

First try to subtract in the ones place. If necessary, regroup only *one* from the next higher place. This *1* gives *10* of what you need and this is always enough (Figure 5.37).

Figure 5.37

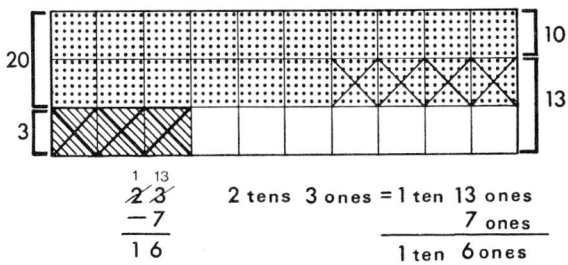

COUNTING ROW ADDITION

Use an 11 square counting strip and find all the pairs that add to the sum of ten. Is there a pattern? See Figure 5.38.

Figure 5.38

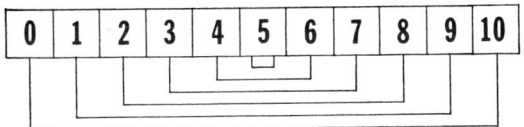

Find all the sets of 3 different addends that add to 10:

0 + 1 + 9 = 10	1 + 2 + 7 = 10	2 + 3 + 5 = 10
0 + 2 + 8 = 10	1 + 3 + 6 = 10	2 + 4 + 4 = 10
0 + 3 + 7 = 10	1 + 4 + 5 = 10	
etc.		

COUNTING CHART ADDITION

How many pairs of addends can you find that add to 50? Examples: 1 + 49; 2 + 48; 41 + 9; 24 + 26. Is there a pattern? Find some sets of 3 addends that add to 50.

Add the 5 consecutive numbers in the inner frame. Find an easier way of determining their total. The first child who discovers the easy way makes a rule and gives it his name: Jim's Rule for Quickly Totaling Five Consecutive Numbers. Have the class select seven consecutive numbers and see if Jim's Rule applies. Does it work for 4? (The middle number is the average. Multiply it by the number of addends.) See Figure 5.39.

Figure 5.39

1	2	3	4	5	6	7	8	9	10
11	12	13	14	15	16	17	18	19	20
21	22	23	24	25	26	27	28	29	30
31	32	33	34	35	36	37	38	39	40
41	42	43	44	45	46	47	48	49	50
51	52	53	54	55	56	57	58	59	60

6.

6

 **Happy Multiplication To You**

—having a multiplication-table struggle? Is long division often "wrong division?" Here's lots of help.

At conference time, do parents ask how they can get into the battle and help banish those multiplication blues?

An excellent topic for a mathematics teacher's speech to a parents group is "How to Help Your Child with His Multiplication Tables." If the usual attendance at the meeting is 50 parents, 200 will show up that evening.

Here is your chance to express your appreciation for all their cooperation and assistance, and to show these interested parents how, by employing the properties of multiplication *naturally* and in *context* with learning the facts, that the task is drastically reduced for everyone, children, parents, and teachers.

Display a large size 100-square grid showing the 1st to 10th multiplication tables. (Or use a similar multiplication chart on the overhead projector.) Here is the essence of your speech:

A quick look at the chart reveals 100 facts to be memorized and retained for quick multiplication computation.

Look again. Further study discloses a way of getting rid of 45

Happy Multiplication To You

of the facts quickly. A staircase, drawn diagonally below the square numbers, divides the chart into two triangular sections, and the same products are repeated in each of the two parts. The reason: if the factors of any selected fact are reversed, a new fact is created but both facts have the same product. The Commutative Law of Multiplication states that this will always be true for any pair of factors. Cover the 45 facts below the staircase, leaving only 55 to go. Criterion for covering these: 3 tests on any complete table with scores of 100% each time. Begin with the 1's.

The Identity Factor for multiplication is 1, for when 1 is used as a factor in any multiplication fact, the product is identical to the other factor. 1 x 2 = 2. 6 x 1 = 6. Compare the 10's table to the 1's table. Both are easily conquered and disappear from the chart.

The 2nd table falls on most children "as a gentle rain from heaven." There were five rolled-up pairs of socks in the drawer this morning. How many socks were there? How many wheels do three bicycles have? Four children in the family — how many hands must be washed before lunch? Suppose there were four sets of twins. Use the Associative Law and the 2nd table twice for these hands. 4 x (2 x 2), or (4 x 2) x 2. Perfect scores and the twos are covered. There are 33 facts left. The 9th table has an intriguing pattern. The finger game is a thinking trick that you will want to pass on to the parents. It's explained in this chapter.

When the 9's are out of the way, only 21 facts show on the chart. Counting nickels helps with the 5th, forks with three tines (could be pitchforks) set the 3rd table, wheels on cars, legs on horses, bears, or tables kick over the 4th. Six difficult facts are left: 6 x 6; 6 x 7; 6 x 8; 7 x 7; 7 x 8, and 8 x 8. For several days each member of the learner's family becomes a special agent and assumes a multiplication fact for a name. "I'm 6 x 6. Say my code name and I'll answer you." For other suitable strategies, see More Helps for Conquering the Multiplication Facts, in this chapter.

Present the parents with small 100-square grids to take home and fill, one for each learner in the family. Suggest the "3-test with 100% on each test" criterion. The child's progress in covering his own chart will show everyone that he is winning the game.

Other highlights of this chapter:

> a special array for presenting the Distributive Law, with explanation of how this property makes multiplication possible beyond the tables, without the learning of further facts

> a workable Commutative Law array and its practical application

- a bulletin board display of multiplication tables that children and teachers construct together
- details on how to present multipliers involving zeros without ending up with rows of useless zeros in the partial products

Careful and intensive study of the versatile multiplication chart brings special rewards. Square number work, factoring activities, distributing with fingers on the chart, finding that the chart is covered with fractions, (see Chapter 8), exploring perimeters of rectangles and how they relate to areas, (see Chapter 7), are valid reasons to nickname this chart, "The Magna Charta."

To maintain the skills and facts that you and your children have spent hours and days achieving, turn to the liveliness of games. Be sure to examine IT'S YOUR TURN, Chapter 9.

Figure 6.1

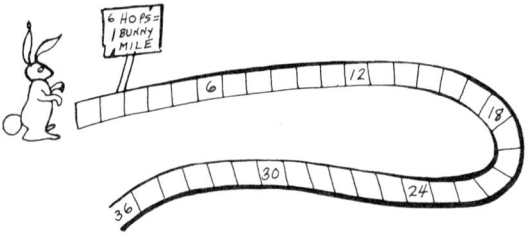

INTRODUCING MULTIPLICATION

1. Equivalent Set Approach

Tape the two sets of squares shown in Figure 6.2 to the chalkboard. Ask children to suggest ways for finding out how many are in both sets. They answer, "Count" or "Add."

Figure 6.2

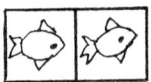

Tape the sets shown in Figure 6.3 to the chalkboard and ask the same question. The answer this time is, "Count," "Add," or "Multiply." By further questioning, elicit the response that multiplication may be performed when the sets are equivalent (have the same number of members), but only counting or addition are used to find the total when the sets are not equivalent.

Record as:

Number of Sets	Number in Each Set	Multiplication Fact
2	2	2 x 2 = 4

Figure 6.3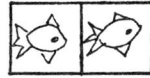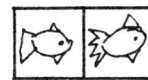

Continue to ask whether or not multiplication may be performed as other sets of squares are formed, and have the reason for the opinion stated each time. See Figures 6.4. and 6.5.

Figure 6.4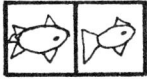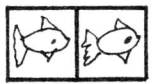

No, these sets are not equivalent.

Add 2 + 2 + 1 = 5

Figure 6.5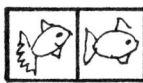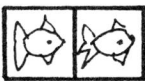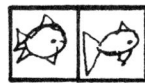

Yes, these sets are equivalent.

2 + 2 + 2 = 6
or 3 x 2 = 6

Continue to display sets of two up to 5 x 2 = 10, have opinions stated for each and facts recorded. Then start removing one set of two at a time until you work back to one set of two squares on the chalkboard. This fact is recorded as 1 x 2 = 2. When the last set has been removed the fact is recorded as 0 x 2 = 0. These two are difficult facts and deserve special attention. Teach "nothing" after teaching "something."

Practice these facts for a few minutes and give a quick test. You will see that the sets approach is a good way to introduce multiplication. Expect everyone in the class to have a perfect score.

2. Repeated Addition Approach

The repeated addition approach is correlated with the set approach. As the sets of squares are taped up, the addition problem is written first and then the multiplication equation (Figure 6.6). See "Multiplication by Repeated Addition of Equivalent Sets" given in detail in this chapter.

Figure 6.6

   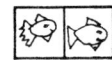 

2 + 2 + 2 + 2 + 2 =   5 x 2 =   10

3. Array Approach

A picture of a parade is shown, and the term "array" is introduced. To have an array all the rows must be equivalent and all the columns must be equivalent. When objects are in an array we can multiply to find "how many." The terms "first factor" and "second factor" are introduced. The array may be turned and the order of the factors changed to introduce the commutative property of multiplication (Figure 6.7). See the parade of the King's Men, Figure 6.29.

Figure 6.7

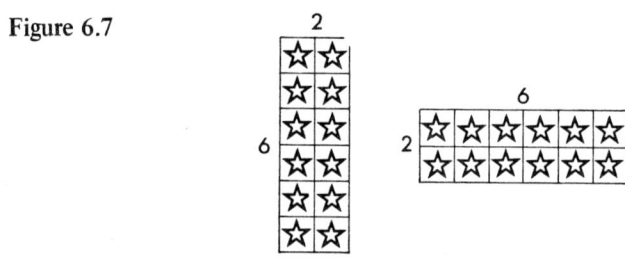

If 6 x 2 = 12, then 2 x 6 = 12

The teacher shows many different arrays and has children write a fact for each. Children make many arrays to illustrate given facts. Be sure they have the chance to make arrays for 1 x 2, which shows only one row of two objects, and of 0 x 2 which shows nothing at all. The term "product" is introduced. Equivalent strips of empty squares are built into a larger and larger array as a particular table is recorded. Arrays are easily produced on squared grids. Paste small pictures, dots, seals, buttons and other small objects within the squares to produce varied and interesting equivalent arrangements.

Arrays lead directly to the study of area of rectangular regions. It is suggested that simple work with area be introduced early in the multiplication unit.

4. Number Line Approach

If a piece of chalk is secretly held behind Jumping Frog, he draws the arcs above a number line as he jumps to illustrate that 8 x 2 = 16. He gives a second performance below the line to show that 2 x 8 also equals 16. Notice that all the jumps look equivalent and the relationship between repeated addition and multiplication is very apparent here.

Figure 6.8

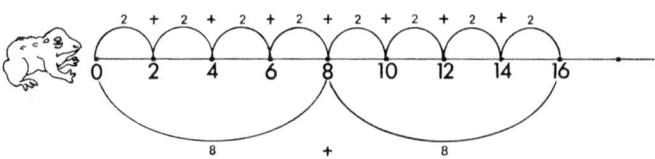

If 8 x 2 = 16 then 2 x 8 = 16. (Figure 6.8)

Happy Multiplication To You

5. Cross Product Approach (Often called Cartesian Product after René Descartes, French philosopher and mathematician.)

Cut 100-square grids to show (Figure 6.9):

Figure 6.9

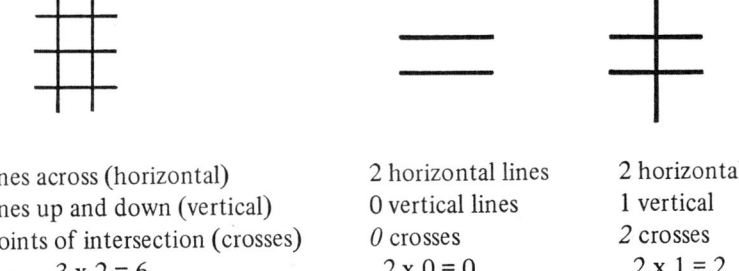

3 lines across (horizontal)	2 horizontal lines	2 horizontal
2 lines up and down (vertical)	0 vertical lines	1 vertical
6 points of intersection (crosses)	*0* crosses	2 crosses
3 x 2 = 6	2 x 0 = 0	2 x 1 = 2

Cross product arrays are filled in to show how many choices of pie à la mode are possible, if there is a choice of 3 flavors of ice cream (vanilla, chocolate, strawberry), and three kinds of pie (apple, blueberry and peach).

Figure 6.10

	A	B	P
V	VA	VB	VP
C	CA	CB	CP
S	SA	SB	SP

There are 3 x 3 or 9 choices of pie à la mode (Figure 6.10).

A BULLETIN BOARD PROJECT — MULTIPLICATION BY REPEATED ADDITION OF EQUIVALENT SETS

1. Noah's Ark (2nd Table)

Every child is given two squares on which to draw a pair of animals. (Discuss the project so each child will draw different animals.) The teacher will supply the Ark, and as the children arrange their beasts in the parade, each will say his part of the second multiplication table. John will remember 7 x 2, for that was when his elephants were "added on." (Figure 6.11)

2. The Baby Elephant's Trunk (3rd Table)

Each child colors three squares. He annexes his squares to the others to help make baby elephant's trunk grow. He says his part of the third table (Figure 6.12).

Figure 6.11

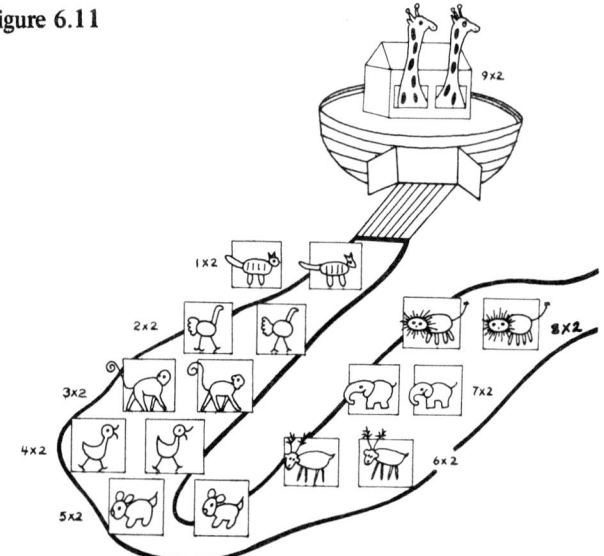

Read loud to the class: "The Elephant's Child," from the *Just So Stories*, by Rudyard Kipling.

Figure 6.12

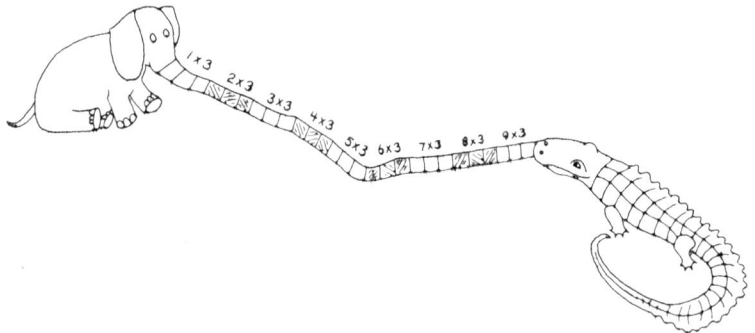

3. The Little Caterpillar Who Wanted to Grow (4th Table)

Each child colors four squares either green or yellow and adds beauty spots. He joins his four squares to help the caterpillar fulfill his desire, and multiplies 4 by the order number of his set (Figure 6.13).

Figure 6.13

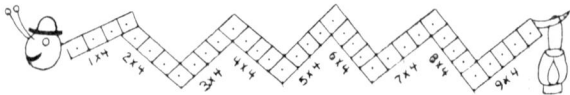

4. The Short Train That Wanted to Be A Long, Long Train (5th Table – Figure 6.14)

Figure 6.14

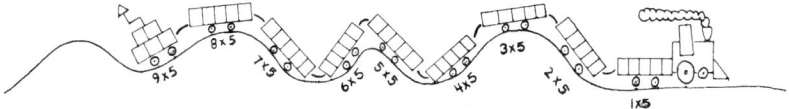

5. The 6-Inch Ruler That Wanted To Be A Carpenter's 6-Foot Ruler. (6th Table – Figure 6.15)

Figure 6.15

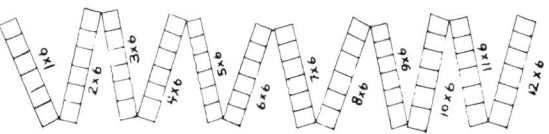

6. The Little House That Wanted To Be A Skyscraper (7th Table – Figure 6.16)

Use large squares and build the skyscraper up the wall.

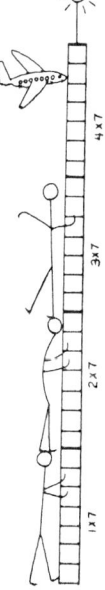

7. The Short Chain That Wanted To Be Long and Strong – Figure 6.17. (The chain is weak if someone doesn't know his 8's.)

Figure 6.17

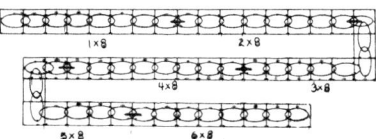

8. **Candles For Christmas.** (Light yours if you can say the 9's.)

Starting several weeks before Christmas, orally test several children each day. Those that can quickly give the correct answers light their bulletin board candles. Those that need more help and study take the test again another day and keep trying until they can light theirs (Figure 6.18).

Figure 6.18

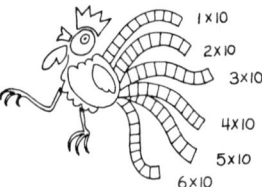

9. Banjo, the Colorful Little Rooster Who Wanted More Feathers In His Tail — so he'd have something to crow about. (10th Table — Figure 6.19)

Figure 6.19

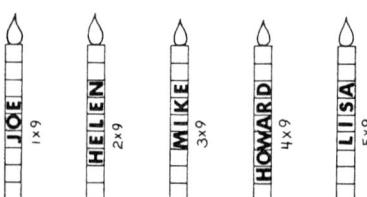

MULTIPLICATION-DIVISION CHART — THE MAGNA CHARTA OF ALL CHARTS (Figure 6.20)

Figure 6.20

X	1	2	3	4	5	6	7	8	9	10
1	1	2	3	4	5	6	7	8	9	10
2	2	4	6	8	10	12	14	16	18	20
3	3	6	9	12	15	18	21	24	27	30
4	4	8	12	16	20	24	28	32	36	40
5	5	10	15	20	25	30	35	40	45	50
6	6	12	18	24	30	36	42	48	54	60
7	7	14	21	28	35	42	49	56	63	70
8	8	16	24	32	40	48	56	64	72	80
9	9	18	27	36	45	54	63	72	81	90
10	10	20	30	40	50	60	70	80	90	100

Happy Multiplication To You

After multiplication has been meaningfully introduced, begin to fill in the compact multiplication chart, and proceed to fill it as facts are learned. There are many methods of constructing the multiplication chart with the pupils. Counting by 1's, by 2's, by 3's, etc. and recording these counts, would produce a multiplication chart.

A technique, that builds *meaning* for the concept of rectangular area and arrays, is using an L-shaped, or right-angle slide. (See Addition Chart, Chapter 5).

The first factors are written down the left side of the 100-square, and the second factors are placed across the top. An example such as 2 x 3 is suggested. The slide is placed around the first factor 2 so there is a "floor" and a "back wall." The slide is moved along the 2 line until the wall is back of the second factor, 3. 6 squares are enclosed within the right angle. The product 6 is written in the angle "corner." The 2 by 3 array can be seen within the right angle. The commutative law for multiplication might be explored by next trying the example 3 x 2, and recording the 6 again in its proper place on the chart. The order of the factors changed but the product remained the same (Figure 6.21).

Figure 6.21

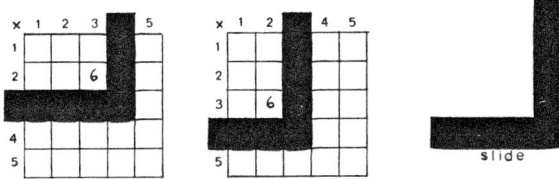

(Pupils work on individual charts with small construction-paper slides as the teacher works with a large grid and suitable size slide. A folding wooden yardstick works well as the teacher's slide.)

A third method would be to have the pupils cut arrays that illustrate different "facts." Each pupil would hold his array against the class chart and record the product on the grid under the last square of the array. He uses the Commutative Law, turns his array and again writes his product on the chart. An area lesson is applicable to this experience with arrays. Covering the multiplication chart with rectangular sets of squares directly suggests measuring a surface with square units, and using multiplication to obtain the total measure. See Chapter 7, Area Is Square Measurement (Figure 6.22).

Figure 6.22

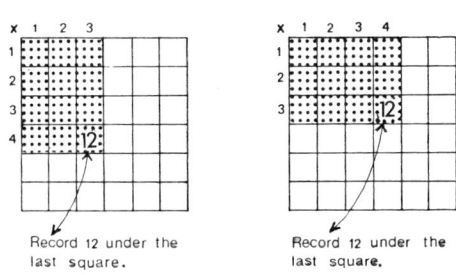

Record 12 under the last square.   Record 12 under the last square.

USING THE MULTIPLICATION CHART

1. For multiplication: Try 2 x 4. The product, 8, is found at the intersection of the 2 row and the 4 column. For extra action, use fingers, the right angle slide, or small cars that run together at the intersection.

2. For division: To find the missing factor for 6 ÷ 3, look for the product 6 in the 3 row, and go up this column to find the missing factor 2 at the top.

3. For division with a remainder or when the quotient is written with a fractional part: using the example, 5 ÷ 2, see that there is no 5 product in the 2 row, but since it is known that 5 comes between 4 and 6, go up the line between the 4 and 6 products to arrive at an answer between 2 and 3 at the top. It will be 2½, or 2, remainder 1.

INTRODUCING DIVISION

Division is the inverse operation of multiplication and to understand this relationship the concept of division should be introduced very soon after multiplication is shown as repeated addition. Division is correspondingly introduced as repeated subtraction. Multiplication may be thought of as putting equivalent sets *altogether*, and division as separating the *altogether* objects back into equivalent sets. If the separation did not produce equivalent sets then the implied idea was of subtraction only and not division.

Show 6 squares altogether and have a child cut them into sets so that each set has 2 squares in it. 6 ÷ 2 = ? The child will be subtracting off equivalent subsets of 2 until the 6 has been divided. He will end up with 3 equivalent sets. He now sees a multiplication problem. He can say 6 ÷ 2 = 3 because 3 x 2 = 6. It is important that the child see this simple idea visualized, as all division algorithms have the multiplication step immediately after the division step. (A remainder, less than the desired equivalent set, cannot be multiplied, but must be added on. See Figure 6.23) See Division on the Counting Chart, Chapter 4.

Figure 6.23

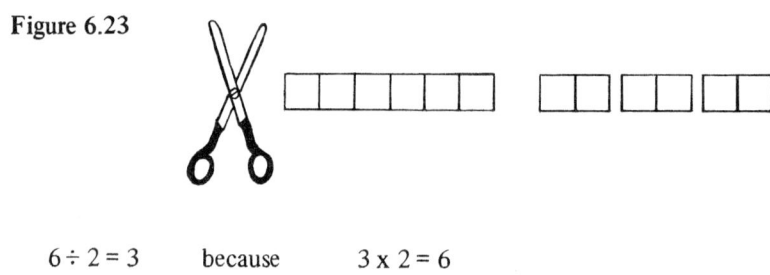

6 ÷ 2 = 3   because   3 x 2 = 6

FINDING AVERAGES – A NEW APPROACH

Take 12 squares, cut three nonequivalent subsets and attach them to the chalkboard. To find the average, have a pupil make the sets equivalent. Equivalent sets mean that the operation of division has been represented (Figure 6.24).

Happy Multiplication To You

When the sets are equivalent, they each have 4 squares. The average of 3, 4, and 5, is 4.

Figure 6.24

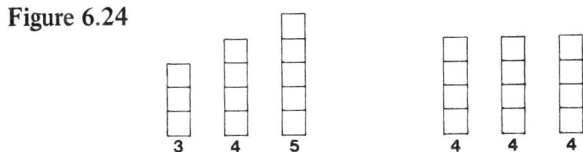

This is fun to do, so have the children find averages by adjusting many different subsets. Try 4, 7, 8. Try 2, 3, 4, and 5. Fractions are often necessary to even things out. As the averaging game becomes increasingly difficult, some deep thinking goes into a search for a better way.

Usually it is when adjusting a set of 17 and a set of 1, in order to make these sets equivalent, that the discovery is made. It is easier first to join the sets and then separate them into two subsets of 9 each. A child explains: "First you add, then divide by 2." (Figure 6.25)

$$\frac{17+1}{2} = 9$$

If the children have been introduced to division as separating, cutting, splitting, sectioning or partitioning objects that are altogether in one set, into *equivalent* subsets, they will realize that the way the 12 squares in Example 1 were separated represents subtraction, but not division. Thus, finding an average is truly dividing.

Story Problem

Pretend you are a bowler. For three games your scores are 98, 102, and 124. What is your average for the three games? Hint: When finding an average it helps to join the nonequivalent sets together and then divide to make the sets equivalent.

Figure 6.25

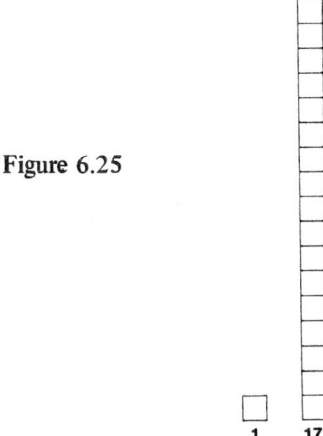

MULTIPLICATION STORY PROBLEM

Pretend you have a fairy godmother who says: "Change your three nickels into pennies and I'll change every penny into a dime." How much money will you have then? (Figure 6.26)

Figure 6.26

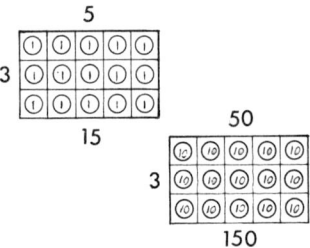

EXPLORING DIVISION

To explore division by 4, give pupils many strips of 4 squares to fill. For 14 ÷ 4, they count out 14 buttons, fill 3 strips, and put 2 on the next strip, which fills ½ of it (Figure 6.27).

14 ÷ 4 = 3 r2, or 3½

Figure 6.27

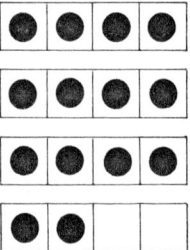

NEVER DIVIDE BY ZERO

Consult a multiplication chart which includes zero factors to discover why zero is never used as a divisor.

Try 4 ÷ 0 = n, or $\frac{4}{0}$ = n. You cannot find the product 4 in the zero row, nor any two numerals between which 4 might be located, so you cannot work the problem. The related multiplication problem is n x 0 = 4, but there is no number for *n* so that n x 0 will equal 4. Any factor times 0 equals 0.

Is it possible to obtain an answer for 0 ÷ 0 = n? There are many 0 products in the 0 factor row. Choose the 0 product under the 3 column and see that 0 ÷ 0 = 3. Choose the 0 product under the 4 column and find that 0 ÷ 0 = 4. Again choose the 0 product under the 0 column and now 0 ÷ 0 = 0. It is impossible to obtain a unique answer for the problem 0 ÷ 0 = n, since there are too many answers. Division by zero is undefined (Figure 6.28).

Happy Multiplication To You

Figure 6.28

x	0	1	2	3	4
0	0	0	0	0	0
1	0				
2	0				
3	0				
4	0				

MORE HELPS FOR CONQUERING THE MULTIPLICATION FACTS

After the concept of Multiplication is well understood and the facts meaningfully introduced, the attention changes to speeding up the facts-recall-time. The child should be able to work multiplication problems without reference to the chart, arrays, repeatedly adding, or using other time consuming aids. Some of the suggestions for helping with the tables that were explained in the chapter notes will be reviewed and given in more detail. Covering the chart is especially rewarding. Here the class is zeroing in on the difficult facts and getting a tiresome job done in an organized manner. The secret — by always selecting the easiest ones left to learn, only the last one can be labeled "difficult." The natural laws of multiplication are used for a real purpose — to make the job easier — and children will remember them as important and friendly laws. In every test, the entire class must succeed — 100% correct by 100% of the class — before the chart is covered. It's everyone's battle — class against chart.

1. The biggest offensive measure is the use of the Commutative Law of Multiplication. The multiplication table is symmetrical about the falling diagonal of square numbers and each product above is repeated below. To emphasize further this double feature of matching products, prepare an array of pictures on the same size grid as the Multiplication Chart. The parade of the colorful "King's Men" is a favorite array. They march on the chart to illustrate that 3 x 6 = 18 above the staircase. They march on again, turn and go up a hill, lose not a man, and show that 6 x 3 = 18 below the staircase. By the Commutative Law, if 3 x 6 = 18, then 6 x 3 = 18, and 45 facts below the staircase may be immediately covered. Children will gain two facts each time they learn one (Figure 6.29).

Figure 6.29

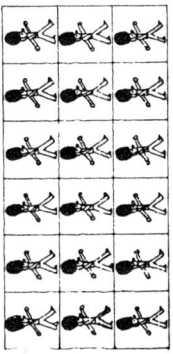

2. There are 55 facts left. What two tables may be readily conquered and covered by the knowledge of what happens when one of the factors in the multiplication problem is 1, the friendly Identity Factor?

3. There are still 36 facts. The second table is usually used to introduce multiplication and is already well known. Also, count by two's and think of ears, shoes, elbows, and sandwiches.

4. Keep selecting the easiest facts from the 28 left uncovered. Do battle with the 9's next. Use the bent finger game. Cut hands of oaktag, tape to the chalkboard and number the fingers from 1 to 10. (A 1 to 10 number line and a pencil may be used in place of fingers.) Figure 6.30, 6.31

Figure 6.30

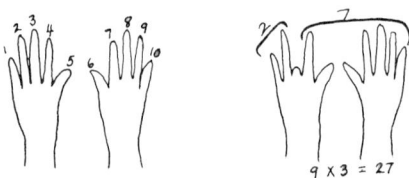

Example:
3 x 9 = ? Bend down the 3rd finger. *2* are up on the left. *7* are up on the right. 3 x 9 = *27*. Try 4 x 9 = ? Bend down 4th finger, leaving up *3* and *6*. 4 x 9 = *36*. Try 5 x 9 = ? Bend down 5th finger, leaving up *4* and *5*. 5 x 9 = *45*.

The pattern is discussed and immediately put to use. Since there are 10 fingers altogether, bending down any one of them leaves up a total of 9. When the 3rd is down, there must be one less than three, or *2* fingers up on the left, and enough to bring the total up to nine, or *7* more up on the right. (2 + 7 = 9) Study the ninth table — the digits of each product add up to 9.

Try 6 x 9 and do not use the fingers. Think, "There must be one less than six, or *5*, up on the left, and *4* more on the right to make nine altogether. Write 54. Notice another pattern of the entire table — the increase by 1 in the ten's place and the decrease by 1 in the one's place. Practice for speed.

Figure 6.31

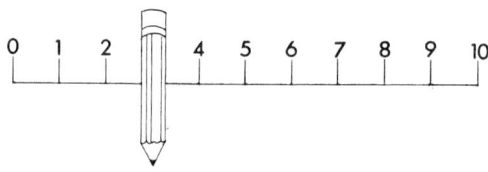

5. There are 21 facts left. Look them over (Figure 6.32).

Figure 6.32

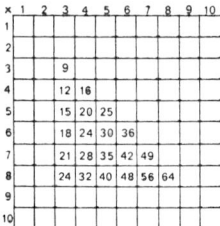

Tackle the 5th table next. Using an "up" and "down" arrangement of squares, count by 5's (Figure 6.33).

Figure 6.33

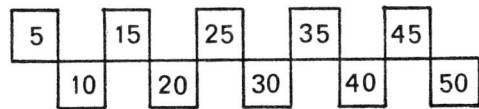

Pattern: 5 x O (odd number) = Up multiple that ends in the odd digit, 5. 5 x 7 = 35.

5 x E (even number) = Down multiple that ends in the even digit, 0. 5 x 4 = 20.

Make a fist of your left hand, and use the knuckles and hollows as ups and downs for learning the 5th table (Figure 6.34). (Have the class pretend they are skipping rope as they count by 5's — softly, on their toes.)

Figure 6.34

Does this strategy work for the 3rd, 7th, and 9th tables? Why?
Example:

3 x 1 = 3   O
3 x 2 = 6   E
3 x 3 = 9   O   etc.

6. Only 17 facts remain. The 3's are next. Make graphs of:

tricycle wheels

clover leaves or shamrocks (or poison ivy if you feel that way about it)

triangle sides

traffic lights

For intensive work use part-charts. Give each child a 3 by 10 array. Have them count the column squares by 3's and record the successive totals in the third row — end up with the 3rd table. Children practice by covering the answers with a strip of paper, and move this along to uncover and verify each answer in turn. In figure 6.35, the next problem is 3 x 5. Moving the slide uncovers the product 15.

Figure 6.35

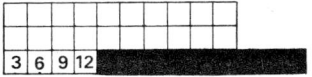

7. Eleven are left and they look lonesome. Now for the 4th table. Make a graph of wheels and cars. See The Visual Impact of Graphs, Chapter 3. Boys like to pick up toy cars one at a time and see how many wheels they can total by multiplication before they miss.

8. Six facts have successfully resisted the onslaught — but not for long.

Arrange 5 rows of desks across the room. A 10-strip of squares is given to each team. Each player fills in a multiple of 6 as the strip is relayed across the room, and fills in another multiple of 6 as it is passed back. The team with their strip correctly filled and back home first wins the race.

9. Only *3* facts remain, 7 x 7, 7 x 8 and 8 x 8. Let children select the easiest one as their next foe and let them suggest and set up a winning campaign. If they wish to battle all three at once, a simple card game race is particularly effective. Only about 10 flashcards are used for the entire race so the three, 7 x 7, 7 x 8 and 8 x 8 will come up often. They will each be worth 5 points and each of the others will have a value of 1 point. A long racetrack of squares will be stretched across the entire chalkboard, the class will be divided into two teams, the X's and Z's. Shuffle the 10 game cards (flashcards are fine), show a card to a member of each team in turn and score 1 or 5 points if answered correctly. No points for incorrect answers. The X and Z move down the track towards the goal (Figure 6.36).

Figure 6.36

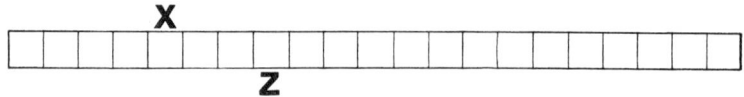

The battle is over when the chart is completely covered. Your children are the victors. Congratulations and awards are in order.

Warning: Continue to practice — never, never stop. Keep working for *retention* and *speed*. See Games, Chapter 9.

DISCOVER WHY THE DISTRIBUTIVE PROPERTY IS THE GREAT LAW OF MULTIPLICATION

The great law of the operations is the Distributive Property of Multiplication over Addition. Why? This law is vitally necessary if multiplication is to be continued beyond the tenth table. When children have learned all their tables through 10 x 10, that's all the multiplication they can do, unless they wish to learn more tables or find a way to distribute the work. They elect the latter and want to know how this works.

Use an array of 3 x 14 squares. Self-adhesive dots are quickly stuck on for color (Figure 6.37).

Happy Multiplication To You

Figure 6.37

×	1	2	3	4	5	6	7	8	9	10 ↓fold				
1	•	•	•	•	•	•	•	•	•	•	•	•	•	•
2	•	•	•	•	•	•	•	•	•	•	•	•	•	•
3	•	•	•	•	•	•	•	•	•	•	•	•	•	•
4	4	8	12	16	20	24	28	32	36	40				
5	5	10	15	20	25	30	35	40	45	50				
6	6	12	18	24	30	36	42	48	54	60				
7	7	14	21	28	35	42	49	56	63	70				
8	8	16	24	32	40	48	56	64	72	80				

Place the array on the Multiplication Chart. It goes beyond the chart and the children realize they have not learned a fact for 3 x 14. (The work must be distributed just as chores are distributed at home — one washes the dishes, another dries). Fold the array into parts that the children do know. They know 3 x 10 = 30 and this fits on the chart. They also know 3 x 4 = 12 and this fits the chart. Addition totals the partial products 30 + 12 = 42.

Write: 3 x 14 = (3 x 10) + (3 x 4)
           =   30   +   12
           =   42

What a wonderful law! Without it we would have to learn the 14th, 29th, 47th and 196th tables and would never come to the end of learning tables. Teaching the natural properties not as separate entities but in context with their usefulness, emphasizes the powerful contribution they make to the mathematical system.

INTRODUCING THE MULTIPLICATION ALGORITHM

Children appreciate shortcuts. Three different ways of recording the distributed work are shown. Use squared paper to emphasize place value and show the break (Figure 6.38).

Figure 6.38

MAXI
```
3 × 14 = 3 × (10 + 4)
       = (3 × 10) + (3 × 4)
       =   30   +   12
       =   42
```

MIDI
```
    1 4
×     3
    1 2
  3 0
  4 2
```

MINI
```
    1 4
×     3
  4 2
```

In all examples, ten ones have been renamed as 1 ten and 2 ones, and added after the multiplication. 3 x 1 ten = 3 tens. 3 tens + 1 more ten = 4 tens.

FINGERS WALK ON THE M-D CHART TO ILLUSTRATE THE DISTRIBUTIVE LAW

1. Use two fingers to illustrate 3 x 14 on the Multiplication Chart. Move the fingers down Columns 10 and 4 to Row 3. The fingers will be on the partial products 30 and 12. Add to get 42 (Figure 6.39).

Figure 6.39

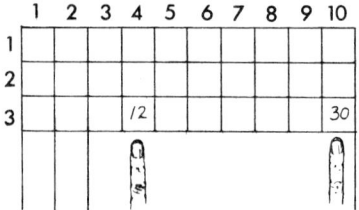

Try (3 x 7) + (3 x 7). Is the result the same? Which is the easier way? Why? It is easier to see the break between the tens and ones places. (1⋮4) Try 3 x 15; 15 x 3; 15 x 9.

2. Use two fingers to work out the entire 12th table. Move down Columns 10 and 2 and add the partial products. Do you see the 12th table under Column 1 and 2? Explain what the 1 represents? The 2? Your fingers will move to 24, 36, 48, 5 and 10 (renamed as 60); 6 and 12 (renamed as 72), et cetera.

See the 123rd table under Columns 1, 2, and 3. Think of Column 1 as 100's; Column 2 as 10's and Column 3 as 1's.

Remember: All these manipulative shortcuts are possible because of the Distributive Law of Multiplication Over Addition.

ESTIMATING WHEN SOLVING STORY PROBLEMS

1. Each child pretends he is a parent. Shoes are on sale for $5.95 a pair (including tax). You are going to buy your three children each a pair of shoes. How much money should you take to the store?

$5.95 ≈ $ 6.00   3 x $5.95 = (3 x $6.00) - (3 x $0.05)
3 x $6.00 = $18.00   3 x $5.95 = $18.00 - $0.15
                         3 x $5.95 = $17.85

2. Estimate the cost of three records at $3.58 each, using upper and lower limits.

Upper limit   3 x $4.00   Answer is too high.
               3 x $3.58
Lower Limit   3 x $3.00   Answer is too low.

Estimate about half-way between the upper limit ($12.00) and the lower limit (9.00) — about $10.50.

Happy Multiplication To You

INTRODUCING 2-PLACE AND 3-PLACE MULTIPLIERS

41
x23

Here is a 4-part distribution and many teachers will want their children to write in all four partial products. The method described below has been developed for the introduction of two-place multipliers for the following reasons:

1. While working with 1-place multipliers, the children have already learned to distribute the multiplicand and have gone from the partial products method (midi) to the regrouping method (mini).

2. Soon problems similar to 24 x 14956 will be used and children will not want to write in all ten partial products. (Giant midis are no fun.)

3. This method leads directly to working with multipliers containing zeros, and eliminates partial products filled with nothing but zeros.

Introduction: Display the problem on large size squared paper. Twenty-three is the factor called the multiplier. The suffix "er" indicates that the multiplier will be the active factor and do the work. The multiplicand will be the passive factor and be worked on. (Other samples of "er" words: farm"er" – works at farming; teach"er" – works at teaching; work"er" – that's you – works on arithmetic (Figure 6.40).

Figure 6.40

		4	1
	X	2	3
	1	2	3	ones part
	8	2	0	tens part
	9	4	3

The multiplier, 23, is a two-place number and since the work will be distributed, there will be a ones part and a tens part. The 3 in the one's place will work first. 3 x 41 = 123. For convenience in adding, the work of the tens should be placed directly under the ones part. Squared paper keeps the work aligned correctly. Multiplying by a tens number always gives a zero in the one's place in the answer. 20 x 41 = 820. Adding the partial products gives the complete product.

It is important always to consider the multiplier to determine the number of partial products. For example, consider the problem: 24 x 14021. There are five places in the multiplicand but only two places in the multiplier and therefore only two partial products.

Zeros in the multiplier do not produce partial products. Study the examples shown below. In the first problem, the multiplier is 302, or 300 + 2. The distribution will be in two parts – a ones part and a hundreds part. No tens are involved. The ones part of the product is determined by the work of the 2 ones. Two zeros are written in the answer when multiplying by the 3 hundreds. Have children explain the distributed work in the other sample problems shown. See Figure 6.41.

Figure 6.41

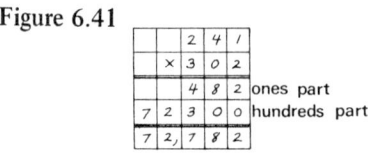

DISTRIBUTIVE PROPERTY PRACTICE

1. Have children draw arrays when given a multiplication fact. If a child finds one particular fact difficult to retain, have him cut an array of large size squares to be placed on the bulletin board. Label it "4 x 6, David." Whenever anyone does not know this particular fact, he refers to David for the answer. (Dots, circles, triangles, or simple pictures may be colored in the squares to make the arrays attractive.) See Figure 6.42.

2. Children cut and color distributive arrays for a bulletin board display. They explain their work to the class.

Figure 6.42

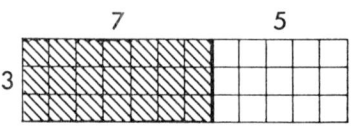

3. Using the value key for a and b, pupils fill in laminated drill cards (Figure 6.43).

Also try:
a^2
b^2
$a^2 + b^2$
$(a + b)^2$ (binomial square)

Figure 6.43

a	10	4	9	3	50	100
b	3	30	9	5	45	90

3a	30				
3b	9				

a+b	13				
3a+b	33				
a+3b	19				

THINK DISTRIBUTIVE – WHEN YOU DO L-O-O-O-O-NG PROBLEMS

To illustrate four-part distributing cut and tape 100-squares to form a large grid that completely covers and extends beyond the multiplication chart, but which may be folded to fit the chart (Figure 6.44).

Figure 6.44

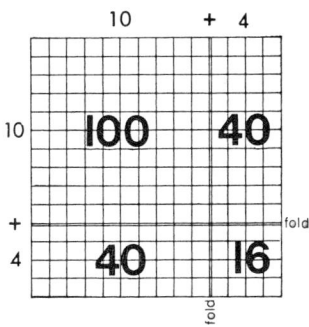

The distribution in four parts:

14^2 or 14 x 14 = (10 + 4) x (10 + 4)
= (10 x 10) + (10 x 4) + (4 x 10) + (4 x 4)
= 100 + 40 + 40 + 16
= 196

Make a face: This round man may be a square (Figure 6.45).

Figure 6.45

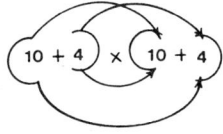

SQUARING A BINOMIAL

$(a + b)^2 = a^2 + 2ab + b^2$

Compare the distributive chart, Figure 6.44, with the chart below. Does it suggest a method for squaring a binomial? Draw one chart on the back of the other to analyze and further emphasize the multiplication similarities of both.

Figure 6.46

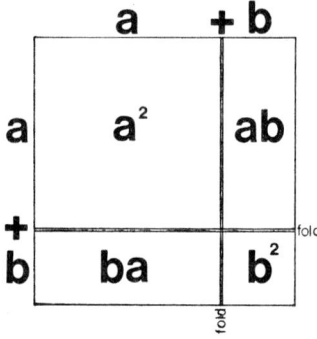

$(a + b)(a + b) = (a \times a) + (a \times b) + (b \times a) + (b \times b)$
$= a^2 + ab + ba + b^2$
$= a^2 + 2ab + b^2$

Discuss Figure 6.47 in terms of $(10 - 4)^2$, or $(a - b)^2$

One explanation:

$(10 - 4)^2 = (10 - 4)(10 - 4)$
$= (10 \times 10) + (10 \times -4) + (-4 \times 10) + (-4 \times -4)$

or $= 100 + (-40) + (-40 + 16)$
$= 60 + (-24)$
$= 36$

Why must 16 be added back on to one of the -40's?

Figure 6.47

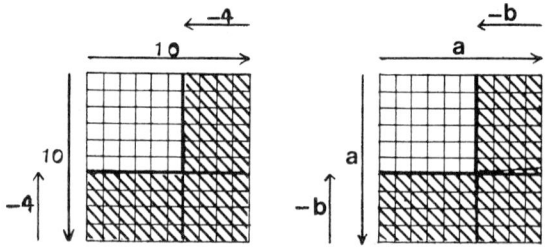

PERCEIVE THE IMPORTANCE OF PLACE VALUE

Multiplication (Figure 6.48)

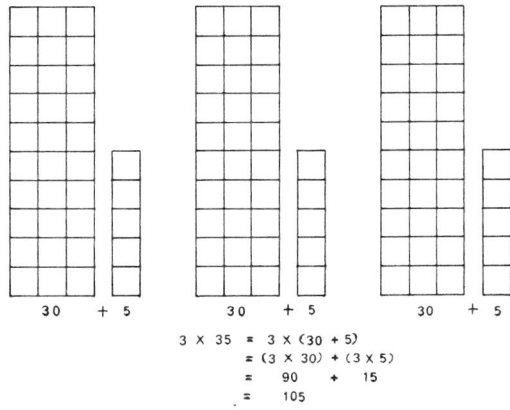

Division (Figure 6.49)

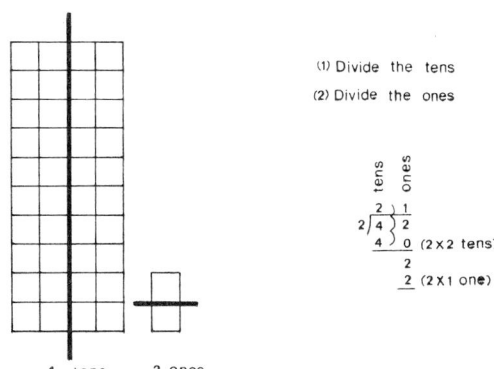

LONG DIVISION – NOT WRONG DIVISION

Squared paper is recommended as a conceptual aid for introducing division by 2-place divisors. As the children manipulate squares of 100's, 10's and 1's, they perform the corresponding number work, writing within large squares to keep their work aligned. Equivalent amounts go to each child.

Story Problem (Figure 6.50):

32 children in the class
353 squares to be divided
3 sheets of 100 squares
5 strips of 10 squares
3 single squares

Figure 6.50

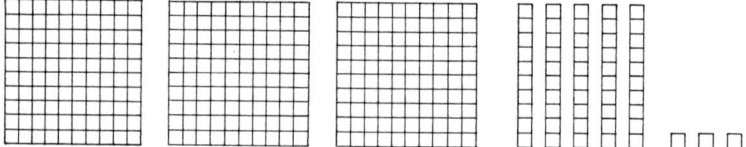

The procedure:

1st   Can you divide the 100's? Place your hand behind the hundreds place and see 3. No, 32 hands are reaching for only 3 sheets.

2nd   Cut the 100's into strips of 10's. Can you divide the 10's? Use your Automatic Changer — placing your hand behind the 10's place automatically changes the 100's to 10's. You see 35 tens. Yes, give each of the 32 hands one strip of 10.

3rd   Multiply 32 x 10 and subtract to see that 33 remain to be divided.

4th   Can you divide the 33 ones? Yes, each of the 32 hands receives 1.

5th   Multiply 32 x 1 and subtract to see that 1 remains to be divided. You can leave this as a remainder, or if you divide it, each hand will get a tiny fraction of this last square. Figure 6.51

Figure 6.51

			1	1	$\frac{1}{32}$	
3	2	3	5	3		
		3	2	0		32 x 10
			3	3		
			3	2		32 x 1
				𝒓		

WINDOWS GIVE A CLEAR VIEW OF
THE ASSOCIATIVE PROPERTY

The Associative Property of Multiplication may be effectively illustrated by windows and window panes constructed of squares. Pictorial proof is obtained by using the same size panes (squares) as the multiplication chart in order to place these panes against the chart in the arrangement indicated by the problem (Figure 6.52).

Example: A. (2 x 3) x 4 may be thought of as a 2 by 3 window seen 4 times, and B. 2 x (3 x 4) may represent 2 windows that are 3 by 4.

Figure 6.52

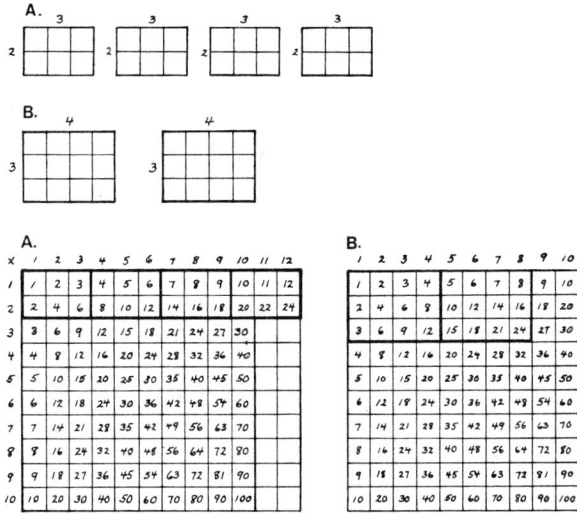

By the Associative Law, if (2 x 3) x 4 = 24, then 2 x (3 x 4) = 24.

A Variety of Bulletin Board Ideas Meaningfully Explain the Associative Law.

Encourage children to explain their thought processes orally as they determine how many window panes are in the houses, petals on the flowers, and squares in the boxes loaded on trucks. Solve each problem in writing, making use of the pictorial representations (Figure 6.53).

Figure 6.53

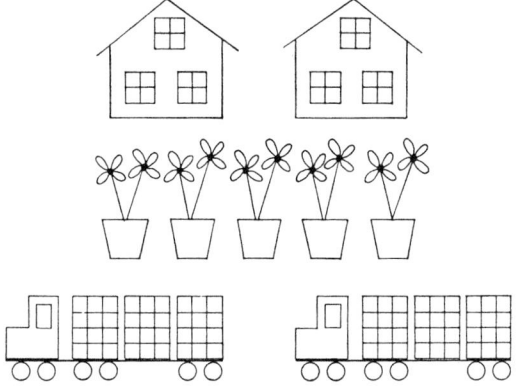

SQUARE NUMBERS AND ONE-OFF-THE-SQUARES

Using grid of similar size, cut and place square arrays on the multiplication chart to show the square numbers and their roots. The square numbers can be located only one place on the chart. Turn the arrays or commute the facts to see why. Brightly coloring the squares containing these numerals calls attention to the symmetry of the chart about the diagonal, makes the chart more attractive, and these "square" facts easier to learn.

The numerals on each side of the squares (off the corners) are one less than the squares. This knowledge provides a trick for learning 3 facts at once. If 7^2 or 7 x 7 = 49, then 6 x 8 = 48, since 6 and 8 are on either side of seven. (6, 7, 8) Also, 8 x 6 = 48.

For enrichment, have higher grade students learn the squares up through the teens, learn how to square the tens and the two-digit numbers ending in 5, and then quickly answer problems similar to these: 15 x 17 = ; 29 x 31 = ; 21 x 19 = ; 13 x 15 = ; 26 x 24 = ; for if 16^2 = 256, then 15 x 17 = 255; if 30^2 = 900, then 29 x 31 = 899, et cetera.

Squaring Tens:

20 x 20 = 400;   30 x 30 = 900;   40 x 40 = 1600;

Squaring Two-Digit Numbers Ending in 5:

Study:

```
   15         25         35         45         55
  x15        x25        x35        x45        x55
  ---        ---        ---        ---        ---
  225        625       1225       2025       3025
   ↑          ↑          ↑          ↑          ↑
  1 x 2      2 x 3      3 x 4      4 x 5      5 x 6
```

The chart in Figure 6.54 shows the geometric squares; their areas, and the relationship of the progression pattern of the odd numbers to the square numbers. (See Chapter 1.) 1 + 3 + 5 + 7 + 9 + 11 + 13 + 15 + 17 + 19 = 100. Throughout the pattern, an odd number of squares are annexed to form the area of the next square number. If the pattern continues to work, the next square number should be 21 + 100. Is this true?

Figure 6.54

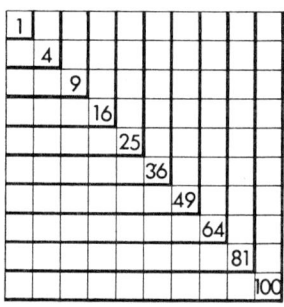

Draw a diagonal through any square number and extend it across the chart. As we go away from the square into the other rectangles, the loss in area is always in progressive odd numbers: 25 - 24 = *1*: 24 - 21 = *3*: 21 - 16 = *5*: 16 - 9 = *7* (Figure 6.55). (See Chapter 7, Building A Pen For The New Pet.)

Figure 6.55

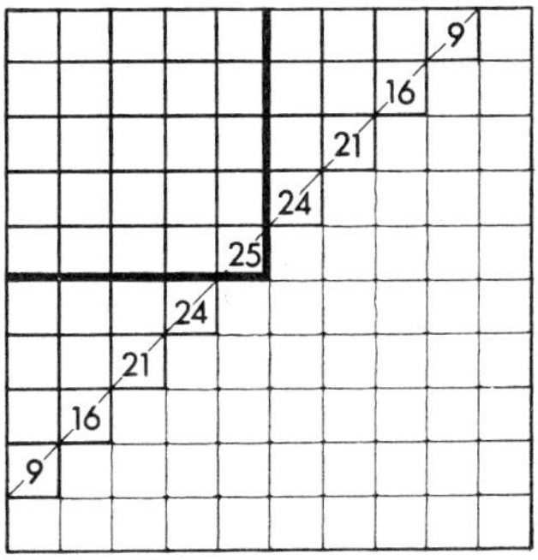

FACTORING ACTIVITIES

Example 1:

Use the multiplication chart to find all the factors of a particular number. First, find the nearest square less than the number, and then test by dividing by only the numbers below the square root. For example, to find all the factors of 48, see that: (1) 36 is the nearest square number less than 48; (2) 6 is the square root of 36; (3) the numbers tested will be from 6 down to 1. (They "catch" the factors above 6.)

Test 6 and get 6 and 8

Test 5 and find that it is not a factor

Test 4 and get 4 and 12

Test 3 and and get 3 and 16

Test 2 and get 2 and 24

Test 1 and get 1 and 48

The set of factors of 48 are [1, 2, 3, 4, 6, 8, 12, 16, 24, 48].

Example 2:

To find the prime factors of 48, test only the prime numbers below the nearest square root: Test 2, Yes; Test 3, Yes; Test 5, No; 2 and 3 are the only prime factors of 48. $48 = 2 \times 2 \times 2 \times 2 \times 3$, or $2^4 \times 3$.

DETERMINING PRIMES AND COMPOSITES

Make as many different arrays as possible for the product 6 (Figure 6.56).

Figure 6.56

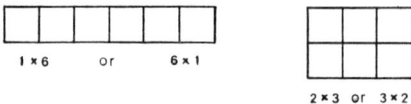

The factors of 6 are 1, 2, 3, and 6.

Make as many arrays as possible for each of the numbers from 2 to 12. Some numbers will have several possible arrays, as 6 has in Figure 6.56, while other numbers will have only one possible array as 5 has in Figure 6.57.

Figure 6.57

Record the results on a chart. Discover that **prime** numbers have only 1 possible array and only 2 factors — themselves **and** one. Composite numbers have more than 1 array and more than 2 factors (Figure 6.58). (When working with arrays no negative integers are involved.) Where can the prime numbers be found on the multiplication chart? Explain. Also, see Chapter 4, Eratosthenes' Sieve.

Figure 6.58

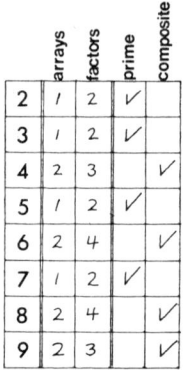

LATTICE MULTIPLICATION (Figure 6.59)

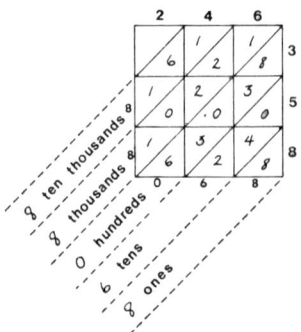

Happy Multiplication To You 139

246 x 358

1. Place factor 246 across the top of the grid, each digit in order above a column.

2. Place factor 358 on the right side of the grid, each digit in order at the end of a row.

3. Draw diagonals across each square in the grid.

4. Multiply 6 (in 246) by 8 (in 358) and place the product 48 in the intersecting square ⁴⁄₈. Multiply 6 (in 246) by 5 (in 358) and place 30 in the intersecting square ³⁄₀. Continue to multiply until all the squares are filled.

5. Add these partial products diagonally to get 8 ones, 6 tens, 20 hundreds (regrouped as 2 thousands and 0 hundreds), 8 thousands, and 8 ten thousands.

6. The product 88,068 is read down the left side and across the bottom of the grid.

7. Check:

```
    246
   x358
   1968    ones partial product (bottom row of lattice)
  12300    tens partial product (middle row of lattice)
  73800    hundreds partial product (top row of lattice)
  88068
```

NAPIER'S BONES

Napier's Bones were originally strips of white wood with the multiplication tables marked on them. Strips cut from the large 100-square grid work equally well. Make a set of "bones" as shown in Figure 6.60. Hang them so they may be easily moved, or glue magnets on back to use on a magnetic board.

Figure 6.60

Multiplier Strip   **Lattice Strips**

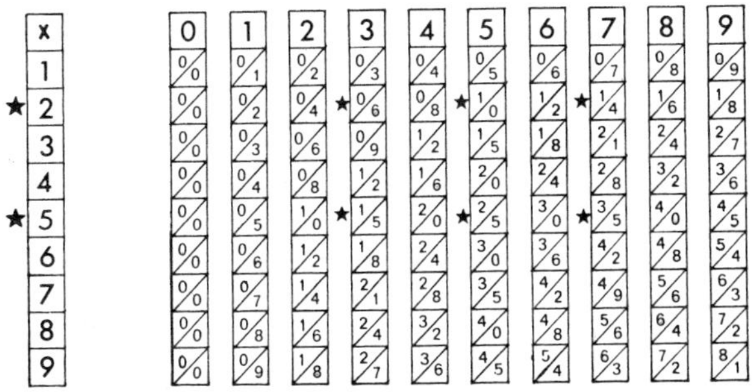

To multiply 5 x 375, place the multiplier strip beside the 3 lattice strip, the 7 lattice strip and the 5 lattice strip, in that order. Use the 5 row on the multiplier strip (Figure 6.61). Add diagonally.

5 x 375 = 1875

Figure 6.61

To multiply 25 x 375, use the 5 multiplier row, and 3, 7, and 5 lattice rows, to gain the ones partial product. Now use the 2 multiplier row and the same lattice rows. Since 2 is in the 10's place, it is really 20, so annex a 0 in the one's place when recording this tens partial product. Add to obtain the complete product (Figure 6.62).

5 x 375 = 1875
20 x 375 = 7500
         9375

Figure 6.62

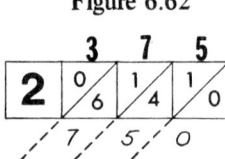

7.

7

 **Measurement**

—a search for the exact

                    If the words "measurement" and "squared paper" were mentioned to a group of teachers, most of them would immediately think of a measurement known as area.

    Children who have been working with multiplication arrays are ready to apply their knowledge to a study of the area of these familiar regions. Their method for finding the number of units in an array, (determining the number of rows and columns, and then multiplying), is the same as that of finding the area of a two-dimensional (2D) rectangular region. In the initial study, many arrays cut from squared paper, some filled with pictures or dots, and some unmarked, are placed to cover the same number of squares on the Multiplication Chart. Areas are directly determined for arrays that represent a tent camp, lunchroom with tables, ranch with cows, parade of the King's guard, window with panes, tiled ceiling, or carpet with unmarked squares. Thus, the connection between area and multiplication is emphasized. The terms "area" and "square unit" are introduced at this time.

    The idea of "covering" a plane region with a number of square units, probably originated when early man measured a plane region by placing his hand flat against the surface as many times as necessary. Children might begin by trying out this hand method, and then move on to covering different areas with square units. Their different hand sizes and the diversity of the squares should lead them

Measurement *143*

to construct standard units — square inch, square foot, square centimeter, and square yard — and to use these for the covering activities. A transparent grid, marked off in square inches, is very useful for quickly covering a small area.

Although governments have established standard square units, teachers should be aware that there are no two-dimensional measuring tools, and that 2D measurement must be accomplished with one-dimensional scales, the foot ruler, yardstick, or meter stick.

These 2D beginnings with squares should lead the children to explore the regional areas of other closed curves, and to the measurement of 1D (one-dimensional or line), and 3-D (three-dimensional or solid) objects. The study of surfaces, edges, perimeter, vertices, angles, prisms and other shapes, similarity, enlargement, congruence, formulas, tessellations, the metric system, working with geometric tools (compass, protractor, T-square), and experimenting with the geoboard are but a few of the supplementary results of squared-paper activity. Tell your children, for they may not be aware of it, that they are "doing geometry."

Some highlights of this chapter are:

Does The Hat Fit? Kindergarteners through Fourth-Graders will love being measured with this height scale.

Feet In A Mile — Little footprints show the number.

Area Is A Square Measurement, and Perimeter Is A Line Measurement.

Big Boys — They provide measurement fun and thought.

Area Cards — Algebraic thinking for the Third-Grade and up.

Sweeping The Sidewalk Reveals Its Area

Building A Play Pen For A New Pet   A study of the relationship of perimeter and area to the shape of the rectangle leaves the child with the lost area in his hand.

Surface Area Of A Prism — Means the area of two bases and four lateral faces

Areas That DO or DON'T Fold Around A Cube

Different Scale Drawings For The Same Dog

A Circle In A Square — Hints for finding the area of a circle

Tessellations — which shapes fit around a point?

Polyominoes — You know dominoes, meet the others

Centigrade To Fahrenheit — and vice-versa — How to do it and to remember how to do it

The Pythagorean Theorem — a square problem

The Square Root Of 2 — Use Squares to locate this elusive number on the number line.

Many teachers tell imaginative stories to enhance their lessons. This is my tale about a ruler.

Man first used his feet, heel to toe, to walk off the length of a room. But when he desired to measure the table top, he did not want to take off his muddy boots and place them, heel to toe, along his nice piece of furniture.

Smart Man! He cut a stick the same length as his foot and placed it, end to end, along the length of the table. He counted, "One foot-stick, two foot-sticks, ... ," or just "One foot, two feet, ..."

This was a great idea, and soon everyone cut a foot-stick of his own with which to measure.

But there was one flaw to correct. The sticks were all of different sizes. Some were long, and some were short, and some were middling. Everyone had cut his to match the length of his own foot, and everyone claimed his to be the correct length. Whose stick do you think was selected as the standard stick for all the kingdom to use? The King's, of course! Since he was the *ruler* of the land, this measuring stick is called the "ruler."

Figure 7.1

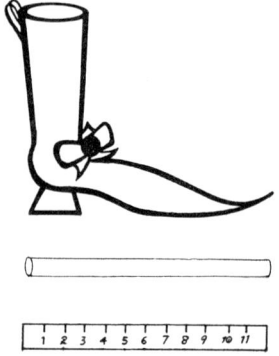

DOES THE HAT FIT?

Build a scale for measuring your pupils' heights (Figure 7.2).

1. Use the large 2" x 2" squares.
2. Cut 20 and 20 and 20.

Figure 7.2

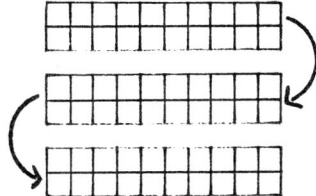

3. Tape ends together and turn lengthwise to form two columns. (See arrows in Figure 7.2)

4. Number right column by 2's from 0 to 60.

5. Mark off inch lines on left column.

6. Tape on wall. Be sure bottom end of scale touches floor.

7. Create big hats of construction paper. (10½" at widest point and 7½" at highest point.) Draw hatband on boys' and cut bright flowers and paste on girls'.

8. Laminate and tape to wall on each side of the scale.

9. To be measured, girls stand under the flowery creation, and boys under the derby (Figure 7.3). Design new hats for each new term.

10. Measure children on the first day of school and again on the last day so they can see how much they have grown during the year.

Figure 7.3

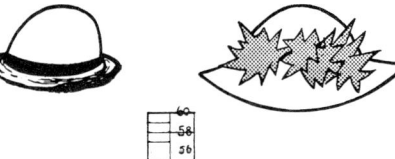

FEET IN A MILE

Many of the measurement conversions are difficult to remember. The little footprints are eye-catchers and memory-strengtheners (Figure 7.4).

Figure 7.4

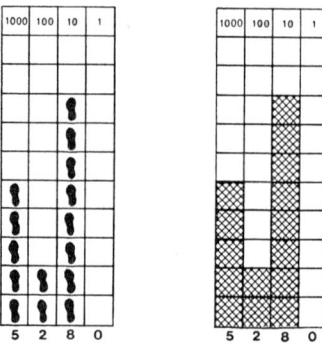

CUPS TO GALLONS

Comparison of units (Figure 7.5)

Figure 7.5

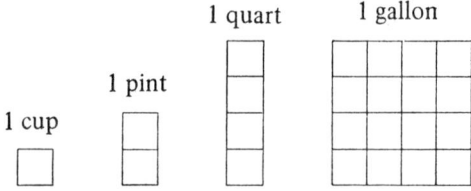

AREA IS SQUARE MEASUREMENT

Using various colored felt pens, fence in rectangular regions of different sizes. Pupils who have been introduced to arrays, while exploring multiplication, will see that a multiplication equation will help determine the number of squares within each closed curve (Figure 7.6).

Thus they draw on previous experience to solve these problems, and learn that the measure of the interior of a plane region is called "area," and the units of measure are "square units." (Figure 7.7) Naming the long side the "length" and the short side of the "width" is readily acceptable to the children.

Children write:

(A)
$5 \times 2 = 10$ $l \times w = A$
Area = 10 square units

Figure 7.6

Figure 7.7

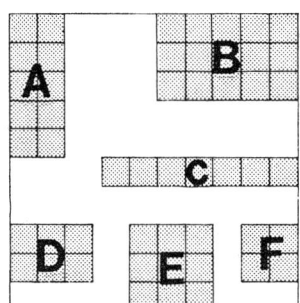
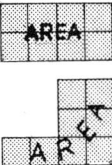

PERIMETER IS LINEAR OR LINE MEASUREMENT

The line around the area is its boundary or perimeter. The measure of this line is also called the perimeter. Look at A in the chart in Figure 7.7. The line around this region forms a rectangle. Count the "line units" all the way around Rectangle A — 2 at the top, 5 on the left side, 2 at the bottom, and 5 on the right side, make 14 altogether. The perimeter of A is 14 units. Find the perimeters for the other rectangles on the chart. Do the same for the rectangles below (Figure 7.8).

Figure 7.8

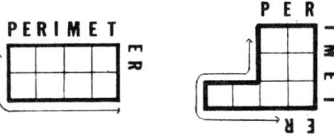

THE CLIMBING INCHWORM

A little inchworm wants to eat a flower sweet that grows at the top of a stalk.

Fifteen inches would be a fine trip if he did not slip two inches each night out of everyday's three inch walk (Figure 7.9).

How many days before he reaches his prize? Be wise! It's a surprise!*

Trips for Inchworm (Figures 7.10, 7.11):
Around:

1. the *rim* of the square inch to measure its pe*rim*eter

2. a square centimeter to measure its perimeter

3. the 4 sq. in. square to find its perimeter

*It took Inchworm 13 days to get to the tulip.

Peri means *around* and *meter* means *measure*. Hint: *rim* is in pe*rim*eter.

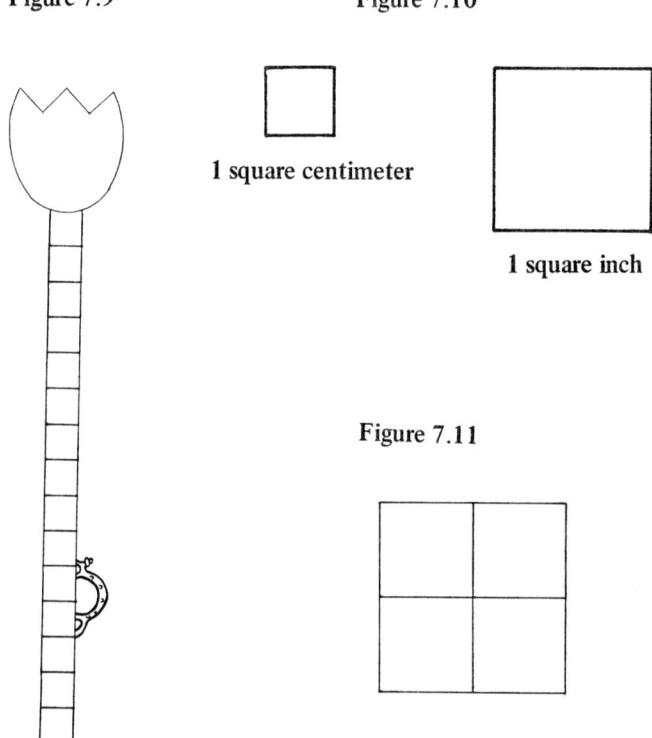

Figure 7.9

Figure 7.10

1 square centimeter

1 square inch

Figure 7.11

LONGER TRIPS FOR INCHWORM

Around:

1. a 3" by 4" index card
2. a 4" x 6" index card

After a friend (that's you) cuts the 4 sq. in. square into four 1 sq. in. squares (Figure 7.12), Inchworm will measure the perimeter of:

3. a stack of four 1 sq. in. squares; a stack of 2; a 3-stack
4. a stack of 10; a stack of 100; a stack of one million (pretend!)

Inchworm flies in his big plane "Firefly" around the house. The dimensions of the house are 30½ feet by 24½ feet. How many feet does he fly? How many inches? See Figure 7.13. (These dimensions allow for a 3-inch clearance for Firefly all the way around. The house is really 20' x 30').

Figure 7.12   Figure 7.13

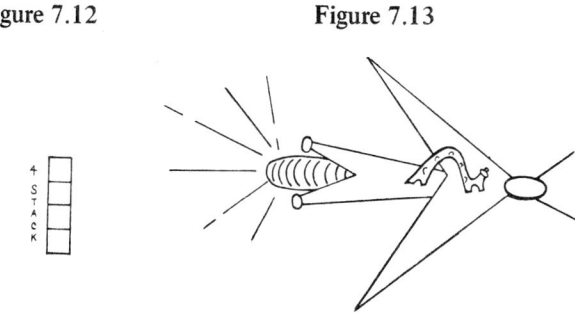

FOUR SQUARE UNITS – MANY ARRANGEMENTS

Give children the opportunity to create areas containing four square units, and be surprised at their resourcefulness (Figure 7.14).

Figure 7.14

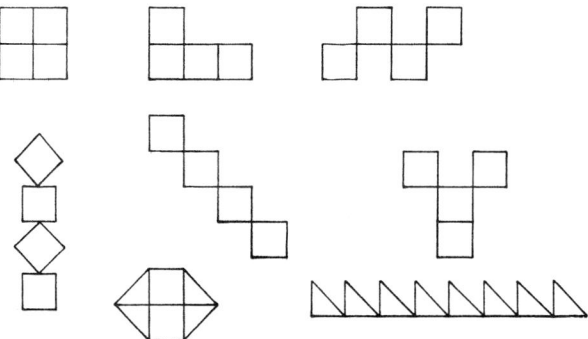

MAKING BIGGER SQUARES OUT OF SMALL SQUARES

One small square unit makes a square array. Does 2? 3? 4? Try up to 25. Keep a record. Find a pattern. Form all the different rectangular arrays you can using 11 square units; 15; 17; 20.

This square activity can be used to develop concepts of area, dimensions, square root, square numbers, primes and composites, multiplication and division facts, commutative law of multiplication, and finding all factors of a number. (See Chapter 6, Multiplication)

STANDARD AREA UNITS

These are models. For the real thing use large grids marked off in square inches (Figures 7.15, 7.16).

1. How many square inches in one square foot? in one square yard?

2. How many square feet in one square yard?

3. Is 144 a square number? What is the square root of 144?

4. $\sqrt{144} = 12$ $\sqrt{144} \times \sqrt{144} = ?$

5. Is 36 a square number? Is 9 a square number? Is 1296 a square number?

6. Are other square numbers illustrated in these charts? How many in the square foot chart? How many in the square yard chart?

Figure 7.15

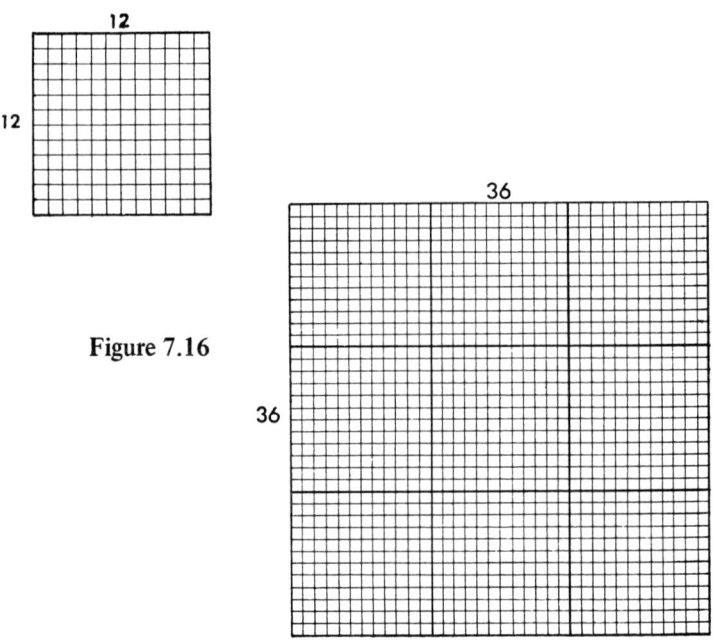

Figure 7.16

BUILDING A PEN FOR THE NEW PET (Figure 7.17)

You have 24 feet of fencing and wish to make a rectangular pen. Look at all the pens shown below. Each requires 24 feet of fencing, but the amount of interior space is different for each pen. Which pen has the greatest area? Which has the least area? Where is the "lost" area?

Find the perimeters and areas of each of the rectangles and record this data in a chart.

Measurement

Figure 7.17

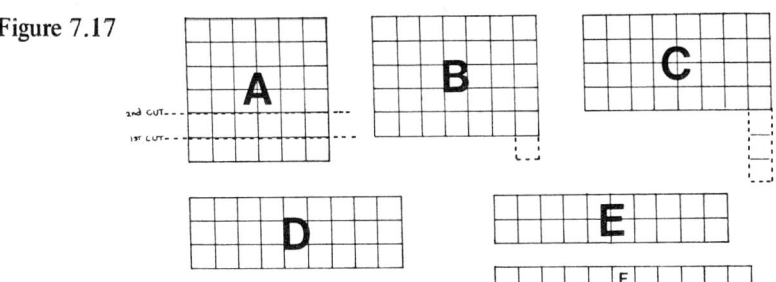

CHART

Rectangle	Dimensions	Perimeter	Area in Square Units	Area Lost
A	6 x 6	24	36	
B	5 x 7	24	35	1 square
C	4 x 8	24	32	3 squares
D	3 x 9	24	27	5 squares
E	2 x 10	24	20	7 squares
F	1 x 11	24	11	9 squares

Change the area of a rectangle, but not its perimeter by following these directions (Figure 7.18):

1. Remove a row from the bottom of A, and attach it back again to the right side of A as a column. This forms a 5 x 7 rectangle and looks like B. One square unit of area is outside the fence. Cut it off.

2. Remove another row from A, and attach it back again to A as a column. A now resembles C. Cut off the "lost" area.

3. Continue this procedure, noting the area each time, until A has one row of 11 units, and looks like F.

4. Record in the chart what happened after each change.

5. Study your chart and answer these questions: Does the area affect the perimeter? Does the perimeter affect the area? Does the shape affect the area?

The experience of the "lost" area is more dramatic if the demonstration is acted out with 36 separate squares having magnetic tape attached, and moved about on a magnetic board. A child removes a row from A, piece by piece, and builds the extra column up the side of A. The one "lost" square remains in his hand where he can actually see and feel it. This procedure of removing rows and annexing columns continues until only one row remains.

Use a string to enclose each rectangle, and thereby show that the perimeter remains constant. Pupils may wish to know what will happen if they continue to lengthen the string and narrow the rectangle. Does the area disappear entirely?

Start with a region of 10 x 10 square units, proceed as before, and see if the loss is always in the pattern of successive odd numbers – 1, 3, 5, 7, 9, 11.

Remember the one-off-the-square numbers on the Multiplication Chart. These are always 1 less than the square number. 6^2 is 36, but 5 x 7 is only 35, or one less than 36, and it appears on the chart on either side of 36. Return to the Multiplication Chart, and examine the numbers that are two-off-the-square. 4 x 8 = 32. Is 32 two spaces away from 36, and one space away from 35? Could the lost area activity with the pet pens be verified on the Multiplication Chart?

Would it be reasonable to assume that a long, ranch-type house would be more expensive than a square or near-square house, provided that the same area and quality of materials were required?

Squared paper is excellent for drawing models and determining the perimeter and area of houses with wings. School houses are often built with several wings. Use squares, and build a model of your own school.

Figure 7.18

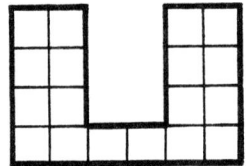

BIG BOYS (Figure 7.19)

Which is bigger, HI BOY or LO BOY?
How do you know? Did you measure

Areas?   Heights?
Perimeters?   Widths?

Tell which is bigger in each case.

Figure 7.19

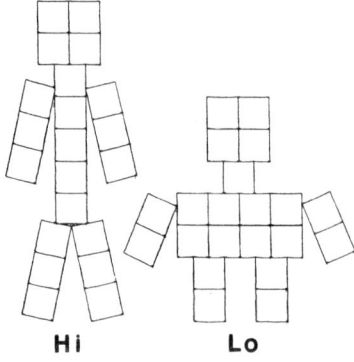

AREA CARDS TEACH ALGEBRA

Prepare these cards with magnets on the back for use on the magnetic board, or in a way that they may be easily handled. Cards are to be held against each other for visual proof, or squares may be counted.

Set 1 (Figure 7.20)

Figure 7.20

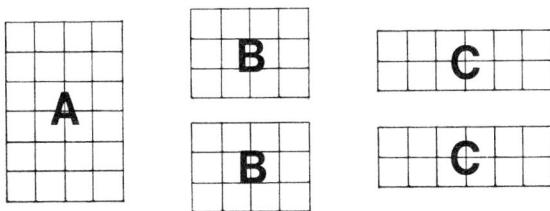

Pupils discover and write expressions comparing area or regions. Generally, algebraic equations are written with lower case letters for variables, as b + c = a or ½a = c.

    _1_ A = _2_ B's   ½ A = _1_ C

    _1_ B = _1_ C   _1_ A = _1_ B and _1_ C

    _1_ A = _2_ C's   _1_ B = ½ A

Set 2 (Figure 7.21)

2 X's = ___ Z's

1 X = ___ Y's

4 Y's = ___ W's

etc.

Figure 7.21

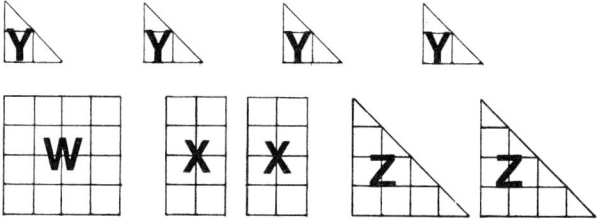

SWEEPING THE SIDEWALK REVEALS ITS AREA (Figure 7.22)

Ken's sidewalk is rectangular in shape and has dimensions 4 feet wide by 10 feet long. You can see how it looks in Figure 7.22 A. Ken bought a snowsweeper as wide as his sidewalk. Everytime he pushes the snowsweeper forward 1 foot, he sweeps 4 square feet. When he pushes it forward 2 feet, he sweeps 2 x 4, or 8 square feet. How many square feet will he sweep if he pushes it 5 feet? All the way?

His friend, Carl, wants to borrow Ken's snowsweeper to sweep his sidewalk. It is pictured and the dimensions given in Figure 7.22 B. As you can see, it curves around several times. Do you think he has more sidewalk to sweep than Ken has? How can you find the area of Carl's sidewalk?

You can count the squares to find the area of Carl's sidewalk, or you can cut it lengthwise through the center, and then piece it together again with the curved edges matching. What is the area of Carl's walk? Did he have more work to do than Ken?

Did you think like this? All those curves should make the work harder than a straight widewalk, but Carl's arms took turns doing the job. At the first curve, his left arm pushed harder while his right arm rested. On the second curve, his right arm worked harder while his left arm rested. So it all averaged out in the end. Each of Ken's arms pushed with the same force — not hard, not easy — just average all the way.

Figure 7.22

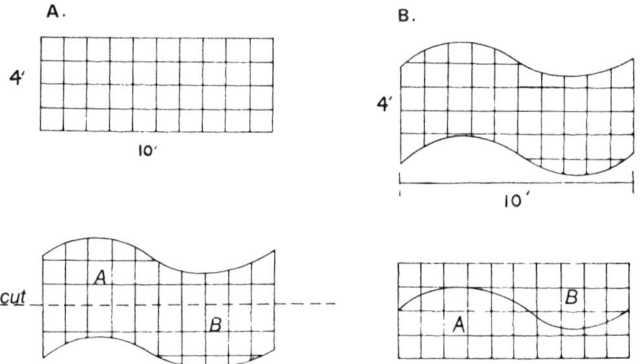

TRIANGULAR AREA – FORMULA DISCOVERY

Squared paper is used to show that a right triangle encloses one half the area of a rectangle (Figure 7.23).

1. Count the square units within the rectangle and those within the triangle. Compare.

2. Fold the rectangular region into two congruent triangular regions.

5 x 5 = 25 sq. units within rectangle

$\dfrac{5 \times 5}{2} = \dfrac{25}{2}$ sq. units within triangle

Formula:

l x w = Area of rectangular region

$\frac{l \times w}{2}$ or ½(l x w) = Area of triangular region

Figure 7.23

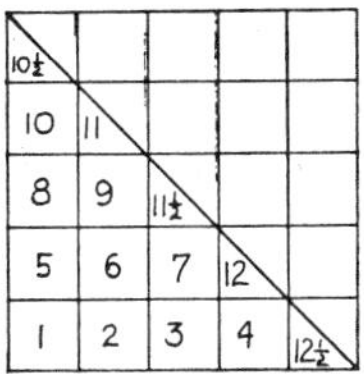

Regions involving other triangles are difficult to fold and count.

1. Match whole squares in the interior and exterior, in a one-to-one number correspondence; then match fractional parts.

2. Cut out the triangle. Piece together the exterior squares to form a congruent triangle (Figure 7.24).

Figure 7.24

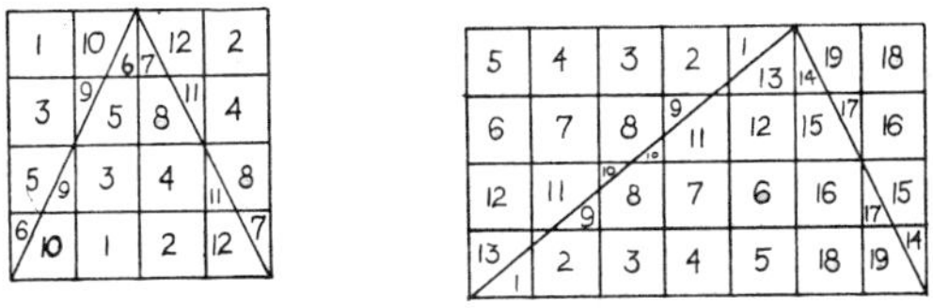

Sweeping (Figure 7.25)
The snowsweeper should be 5 feet wide.

1. Start at the base of the triangle which is 5 feet wide and end at the apex which is 0 feet wide. The average *width* swept is $\frac{5+0}{2} = \frac{5}{2}$ ft.

2. The *length* swept is 5.

3. w x l = $\frac{5}{2}$ x 5 = 12½ feet.

4. Area of triangle: $\frac{w}{2}$ x l

Figure 7.25

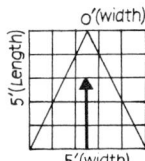

AREAS THAT DO OR DON'T FOLD AROUND A CUBE (Figure 7.26)

A cube has 6 faces. The surface area may be illustrated by 6 squares cut all in one piece. Study the outlined regions. Which ones would fold around a cube? Copy the regions, cut them out and fold them to see if your answers were correct. Try some others. (These 6-square figures are hexominoes. There are 35 different arrangements, 11 of which will fold around a cube. See Polyominoes, this chapter.)

Figure 7.26

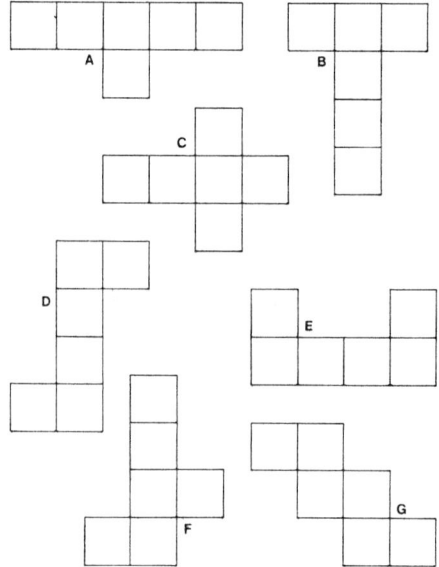

LOST AREA PUZZLE

The 8 x 8 grid has 64 square units. Mark and cut as in Figure 7.27.

Figure 7.27

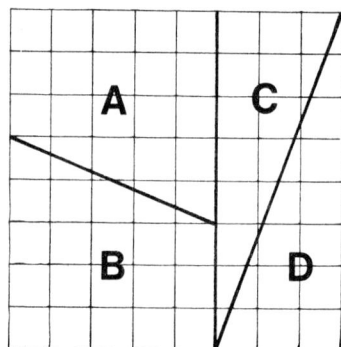

Using the 4 pieces, form a 5 by 13 rectangular array. 5 x 13 = 65. (Figure 7.28) Explain the extra square unit. (The fit is not exact along the diagonal.)

Figure 7.28

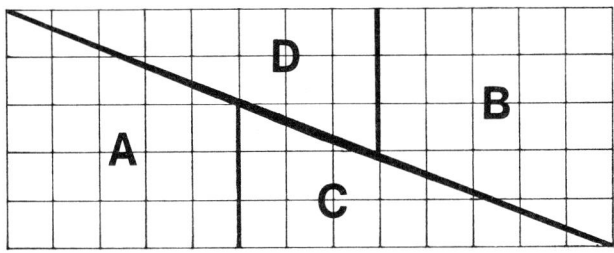

SAME DOG – DIFFERENT SCALE

Scale drawings, small to large, are produced on (1) small squares, divided into 4 square units; (2) small squares; (3) large squares, divided into 4 square units; (4) large squares. The full picture of the dog may be drawn on an entire sheet of the large 100-squares (Figures 7.29, 7.30, 7.31).

Figure 7.29

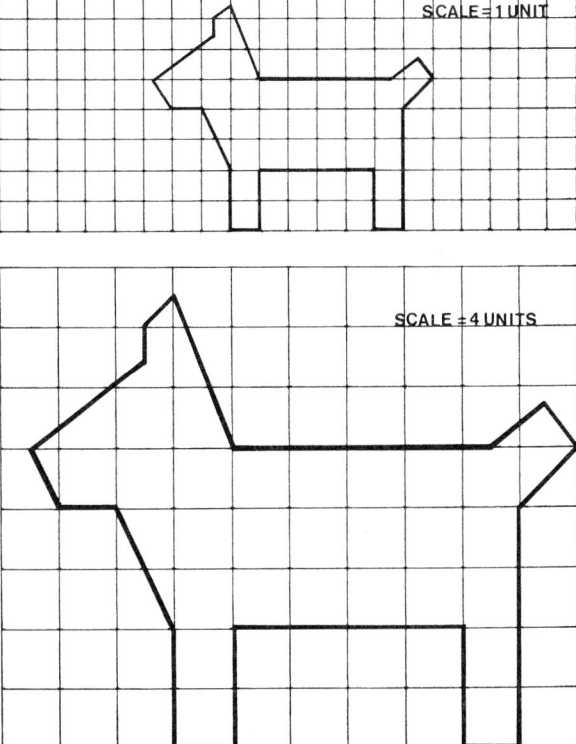

158   *Measurement*

Figure 7.30

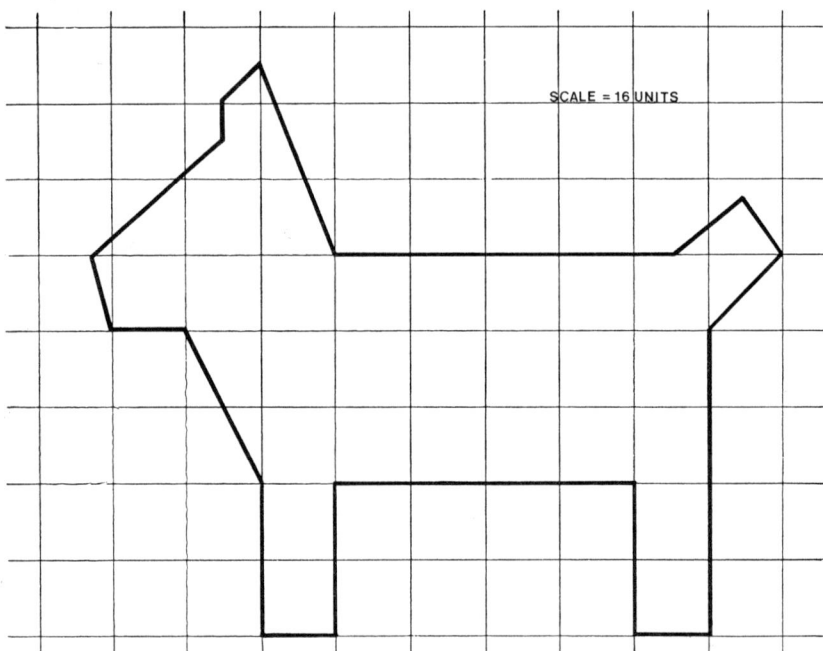

SCALE = 16 UNITS

Figure 7.31

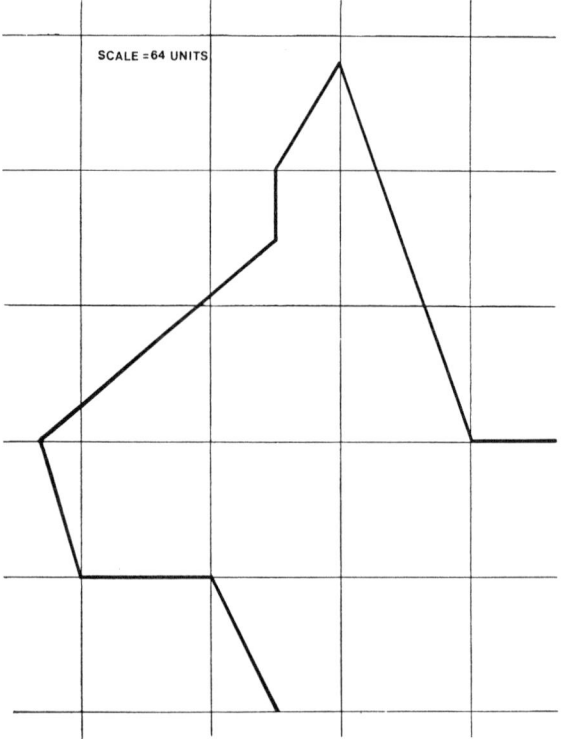

SCALE = 64 UNITS

A CIRCLE IN A SQUARE – HINTS ON FINDING THE AREA AND THE FORMULA (Figure 7.32)

Figure 7.32

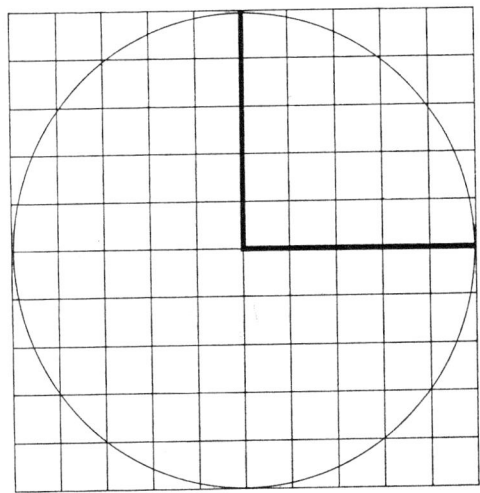

1. Multiply to find the area of the large square. 10 x 10 = 100 sq. units.

2. Count all the squares inside the circle to find the area of the region. Approximately 78 sq. units.

3. The small square is ¼ of the large square. Its area is 5 x 5 = 25 sq. units.

4. The sides of the small square are also the radii of the circle. Thus, r = 5, and $r^2 = 5^2 = 25$ sq. units.

5. $4r^2 = 100$ sq. units. Some of this area lies outside the circle. Only 78 sq. units lie inside the circle, so *4* times r^2 is too great.

6. Try 3 times r^2. $3r^2 = 75$ sq. units, which is less than 78 sq. units.

7. Try Pi times r^2. $\pi r^2 = 3.14 \times 25 = 78$ sq. units (approximately).

8. Formula for finding area of a circle: πr^2

A GIANT PROTRACTOR

A protractor large enough for class use may be constructed on the large 100-square matrix. For stiffness needed for easy handling, and also for durability, lamination is suggested. (Figure 7.33)

Angles, drawn on another sheet of large 100-square paper, are measured by correctly placing the matching-size protractor over them, Figure 7.34. (See Chapter 8, Fractions.)

Figure 7.33   **Figure 7.34**

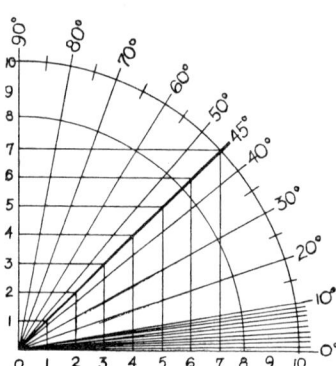
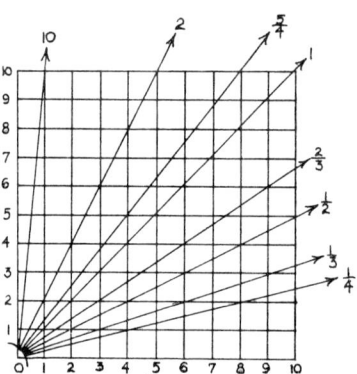

SURFACE AREA OF A PRISM MEANS FINDING AREA OF 2 BASES AND 4 LATERAL FACES

Laboratory work with visual materials help to introduce Surface Area. Pupils are to:

1. Look at a picture of a rectangular prism. (Fig. 7.35)

2. Handle a box, or a wooden prism of equivalent size.

3. Build a model of a 4" x 2" rectangular prism from cubic-inch blocks. (Fig. 7.36) Glue the blocks together for easier handling.

4. Identify the lateral faces and the bases, and find the surface area by any method they devise.

5. Cut square-inch grids to show each of the 6 faces. (Fig. 7.37)

6. Cut a model of the entire surface. (Fig. 7.37)

7. Explain carefully how they arrived at their answers.

One explanation:
There are 4 congruent lateral faces, each with an area of 4 x 2, or 8 square inches: 2 base faces, each with an area 2 x 2, or 4 square inches.

This may be written as:

4 x (4 x 2) + 2 x (2 x 2) =
   (4 x 8) +(2 x 4) =
      32   + 8   = 40 square inches

Measurement

Figure 7.35

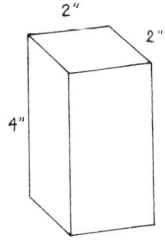

Figure 7.36

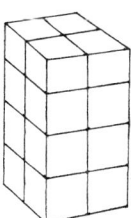

Figure 7.37

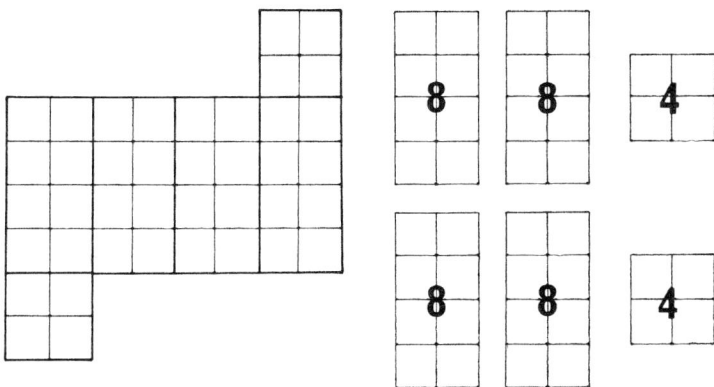

Pretend you own a dog who needs a doghouse. Think out the problem of building him one. Draw a picture. What do you need? How do you measure? Estimate the cost.

TESSELLATIONS ON SQUARED PAPER

A tessella is a small tile. Many shapes are available, round, square, triangular, crescent, T-shaped, H-shaped, L-shaped, pentagonal, hexagonal, octagonal, and so forth.

Designing and coloring tessellations on squared-paper is a rewarding substitute for the arranging of real tiles. Joining the tiles together to form a mosaic pattern is an accomplishment, and embedded in this challenging external experience are internal mathematical learning experiences.

When pupils are asked why squares fit together so nicely, they have a difficult time describing what they intuitively know – squares just do – it's the way their corners are made.

There is a way of telling which regular polygons will tessellate without extra joining shapes. Experiments show that they must fit around a point without overlapping or gapping. Squares do this — four fit around a point or vertex. Since each of these vertex angles measures 90°, four will measure 360°, the measure around a point in a plane. 4 x 90° = 360°, or 360° ÷ 90° = 4 (Figure 7.38).

Figure 7.38

What other shapes fit around a point?

An equilateral triangle is a 3-sided polygon with 60° angles. 360° ÷ 60° = 6, so 6 equilateral triangles will tessellate around a point.

A regular hexagon is a 6-sided polygon with 120° angles. 360° ÷ 120° = 3, so 3 hexagons will fit around a point. A honeycomb is an example of hexagonal tessellation.

Test other regular polygons. Are there any that do not tessellate?

A regular octagon is an 8-sided polygon with 135° angles. 360° ÷ 135° = 2^+ or 3^-, so 2 octagons will leave a gap and 3 will overlap.

Since tessellating the grid will be limited to what shapes will fit squared paper, experiments will rule out many of the regular polygons unless they are used with joiners.

OCTAGONS WITH SQUARE JOINERS (Figure 7.39)

Figure 7.39

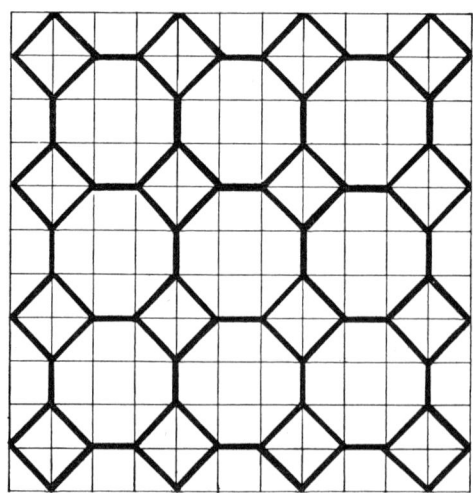

Measurement

POLYOMINOES – YOU KNOW DOMINOES, MEET THE OTHERS

Polyominoes are sets of square arrangements with each square having at least one edge joined to another. The familiar domino has only 2 squares and 1 arrangement. Some other polyominoes are:

Squares	Names	Arrangements
3	Tromino	2
4	Tetromino	5
5	Pentomino	12
6	Hexomino	35

A set of the 12 pentominoes is shown in Figure 7.40.

Figure 7.40

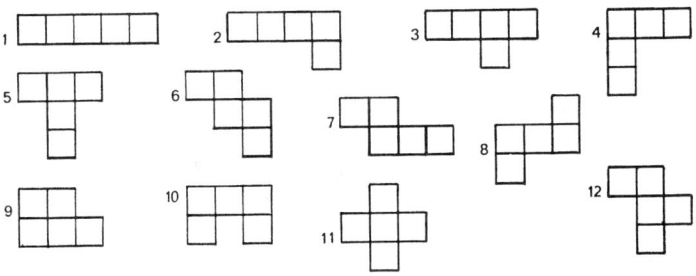

TESSELLATE POLYOMINOES

Select a pentomino and experiment to see if it will tessellate on the squared grid. (Figure 7.41) Also, using each of the twelve pentominoes only once, form a rectangular region. They will fit into a 3 x 20, 4 x 15, 5 x 12, or 6 x 10 rectangle. (Figure 7.42).

Pentominoes (Figure 7.41)

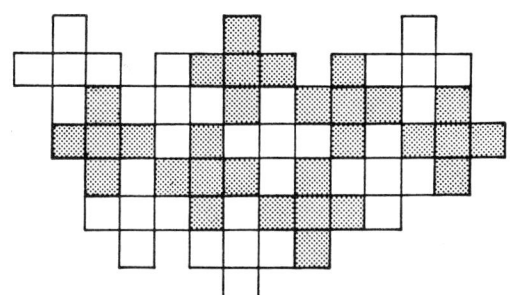

Figure 7.42

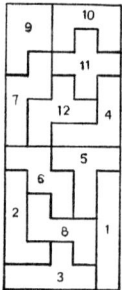

A traveler in Portugal and Spain soon finds his attention focused on the sidewalks. Constructed of small stones or marble tiles, each walk is an example of tessellation. A guide explained that these works were not created solely to please the artistic sense, but were also the air-conditioners of olden times. Water was poured on the sidewalks and the gradual evaporation cooled the passerby as he strolled beneath the trees.

Trominoes

Figure 7.43

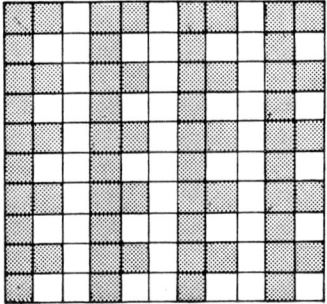

PYTHAGOREAN THEOREM – A SQUARE PROBLEM

Use squared paper to show the relationship of the sides of a right triangle. By the Pythagorean Theorem, side "a" squared, plus side "b" squared, equals side "c" squared, or $a^2 + b^2 = c^2$ (Figure 7.44).

$a^2 + b^2 = c^2$ **Figure 7.44**
$3^2 + 4^2 = 5^2$
$9 + 16 = 25$

$a = \sqrt{9} = 3$
$b = \sqrt{16} = 4$
$c = \sqrt{25} = 5$

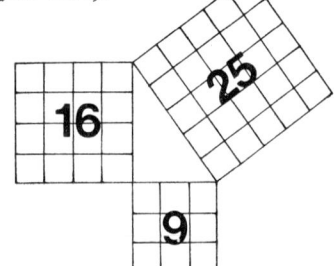

Measurement

THE SQUARE ROOT OF 2 — Use Squares to Locate It on the Number Line

To locate the point for $\sqrt{2}$ on a number line, draw two number lines on squared paper to form a right angle. Mark off the points 1 and 2 on each line, and divide each unit interval into thirds.

Construct a right angle with sides a and b, and hypotenuse c. Since, by the Pythagorean Theorem, $a^2 + b^2 = c^2$, in this particular triangle, $1^2 + 1^2 = 2$, and $c = \sqrt{2}$. Using a compass, describe an arc so that it touches each number line at $\sqrt{2}$. It can now be seen that the $\sqrt{2}$ is between $1\frac{1}{3}$ and $1\frac{1}{2}$, or $1\frac{1}{3} < \sqrt{2} < 1\frac{1}{2}$ (Figure 7.45).

Figure 7.45

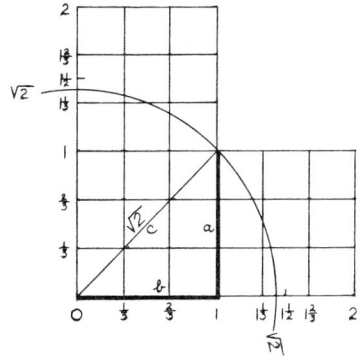

FAHRENHEIT TO CENTIGRADE (Figure 7.46)

Examine the thermometer scales:

1. Centigrade: from freezing point (0°) to boiling point (100°) there is a difference of 100°. One-half the difference is 50°.

2. Fahrenheit: from freezing point (32°) to boiling point (212°), the difference is 180°. One-half the difference is 90°.

3. Every 10° C. = 18° F. or every 5° C. = 9° F.

4. The ratio of Centigrade is Fahrenheit is $\frac{100}{180}$ or $\frac{50}{90}$ or $\frac{5}{9}$

5. The ratio of Fahrenheit to Centigrade is $\frac{180}{100}$ or $\frac{90}{50}$ or $\frac{9}{5}$

6. Formulas for changing scales:

C. = (F. - 32°) x $\frac{5}{9}$ and   F. = (C. x $\frac{9}{5}$) + 32°

166 *Measurement*

Figure 7.46

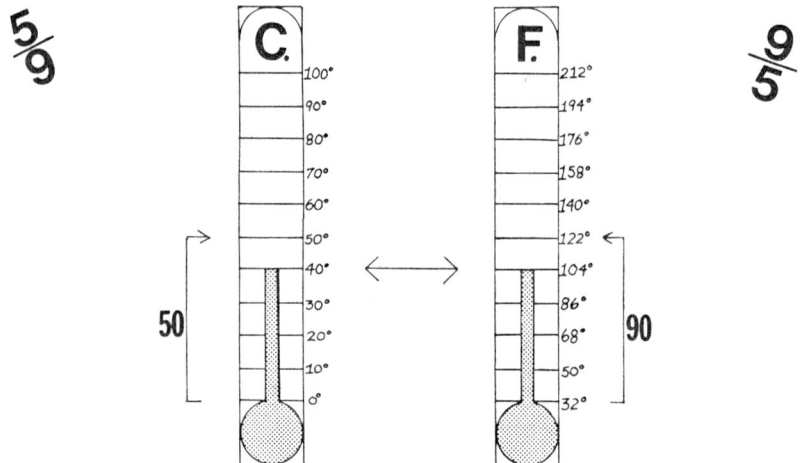

8.

8

 # Fractions - Bits of This and That

—action is a big part of fraction

Certainly, 100 is an ideal number of squares for comprehensive work with the percent ratio. "Per" means "compared to," or simply "to," and "cent" means "hundred." Any part of the 100-square grid is a percentage of the entire grid. Have a child perform the action of running a finger along 15 squares, and say, "I have just touched 15% of the (100) squares." Another pupil, who touches 50 squares remarks that he has touched one-half of the hundred squares, or 50% of them. He would record his action as the numeral 50/100, or 5/10, or 1/2, or in the decimal fraction form of .50, or .5.

The decimal fraction with denominators as powers of 10, may also be effectively displayed on a grid divided 10 to a row and 100 in entirety. Fractional units of one-tenth and one-hundredth are easily perceived. Have a child color .2 and .03 more of the squares. This action produces color in .23 or 23/100 of the chart. The crayon touched 23% of the squares, and this is also .23.

Unit rectangular regions are cut from the large grid and the squares shaded to illustrate any fractional number. The activity of cutting the correct number of squares to represent the denominator, and then coloring the correct number for the numerator, gives meaning to these terms of the fraction and is a valuable learning experience. To show 3/4, cut a unit of 4 squares and shade 3. This conception of the fraction 3/4 leads directly to addition and subtraction of fractional numbers. Other meanings of 3/4 include the ratio of 3 to 4, the designation of 1/4 of each of 3 whole units, and the answer to the division problem $3 \div 4$. The answer, 3/4, also means 3 divided by 4, and so is the problem as well as the answer. Two sandwiches divided by 3 girls means each girl will get 2/3 of a sandwich; 4 apples divided by 5 boys gives each boy 4/5 of an apple.

The concept of equivalence is intuitively discerned when the shaded part of a unit is first given a fractional name, and then the entire unit is further divided. The result: more shaded parts (higher term numerator), more unit parts (higher term denominator), a different fractional name, and yet the area of both the shaded part and the unit region remain the same. A grid for the overhead projector is a convenient way to present equivalent fractions. You see through the problem as you look through the transparencies and colored acetate overlays.

A careful examination of the multiplication chart suggests many equivalent fractions. Inspection of any two rows reveals ten equivalent fractions and this number does not stop when the ends of the rows are reached, for every fraction belongs to an unlimited equivalence class. Graphing fractions is an excellent project. Equivalent fractions stay in their own lines, proper fractions fall below 1 and improper fractions above, and the line for fractions with zero denominators can never be reached. A slide rule checks fractional addition and subtraction problems, and renames fractions automatically.

Rectangular arrays of squares suggest areas, and areas suggest multiplication. If the entire array is named as 1 unit, then any part of the array has an area less than 1, and is named by a fractional number. Multiplication does not always produce a larger number. One-half of one-fourth gives only one-eighth.

Pictures of bright red apples in an array help explain the distributive law of multiplication and its use when solving a problem of a whole number times a mixed number.

Return again to the multiplication chart to find the Greatest Common Divisor (Factor) of two numbers, and use it in conjunction with the counting chart to discover why Euclid's Algorithm for finding the greatest common divisor works so well.

There's fun in this chapter, too, when you weigh a whale – or a snail.

Figure 8.1

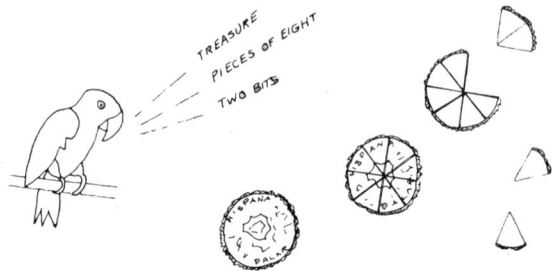

COLOR-SQUARED PROBLEMS TO MAKE YOU THINK

For each set of pictures write an equation that gives the solution to the problem (Figure 8.2).

Figure 8.2

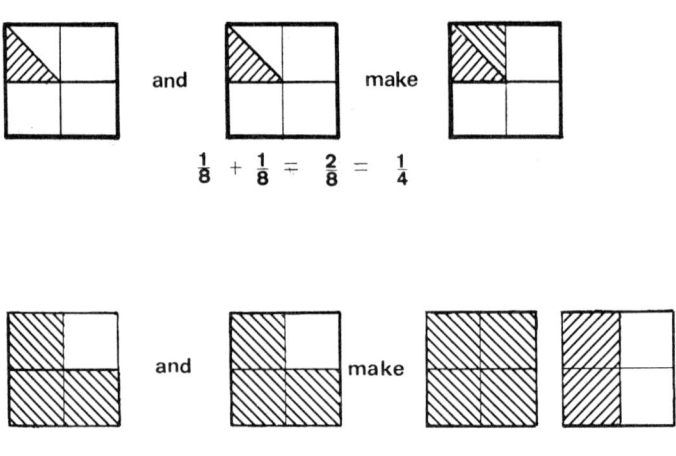

$\frac{1}{8} + \frac{1}{8} = \frac{2}{8} = \frac{1}{4}$

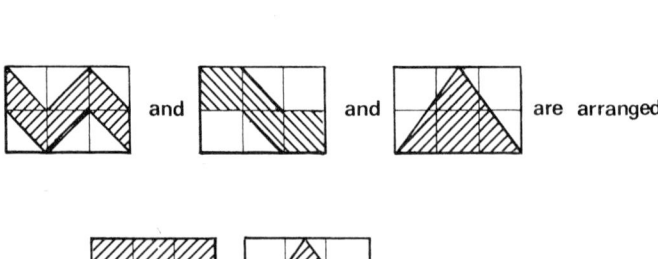

Fractions – Bits of This and That

After children are familiar with this activity, they may be given only the problem and asked to solve it by coloring in the answers themselves.

If pupils fail to recognize the fractional parts, have them cut out and arrange the parts as shown in the answers.

IDENTIFYING FRACTIONS

Supply each pupil with a sheet of hundred-squares, crayons and scissors. Designate the fractional numbers that are to be illustrated. The pupils will soon recognize that the denominator indicates the number of squares they must cut as one unit, and the numerator indicates the number of squares they should color, and if they cut twice as many for the denominator they must color twice as many for the numerator (Figure 8.3).

Figure 8.3

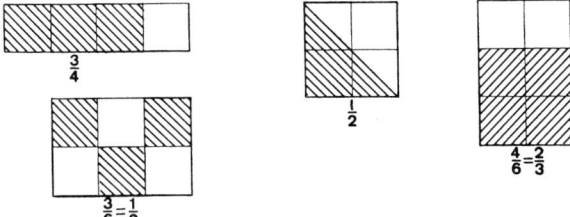

CONSTRUCT A FRACTIONAL ADDITION-SUBTRACTION CHART

Squared charts enable the learners to discover that addition of fractions involves adding the numerators (number of fractional parts), and is similar to addition of whole numbers (number of units). See Figure 8.4.

1. $\dfrac{2}{6} + \dfrac{1}{6} = \dfrac{2+1}{6} = \dfrac{3}{6} = \dfrac{1}{2}$ 2. $\dfrac{2}{6} + n = \dfrac{4}{6}$ $n = \dfrac{2}{6} = \dfrac{1}{3}$

3. $\dfrac{2}{8} + \dfrac{1}{8} + \dfrac{4}{8} = \dfrac{2+1+4}{8} = \dfrac{7}{8}$

Figure 8.4

+	1	2	3	4
1	2	3	4	5
2	3	4	5	6
3	4	5	6	7
4	5	6	7	8

+	1/8	2/8	3/8	4/8
1/8	2/8	3/8	4/8	5/8
2/8	3/8	4/8	5/8	6/8
3/8	4/8	5/8	6/8	7/8
4/8	5/8	6/8	7/8	8/8

FRACTIONS OF SETS

Introduction with real objects — 3 books on the table — Phyllis took 2. She took 2/3 of the set, and left 1/3. Follow up with semi-concrete picture chart. (Fig. 8.5) Variation: Each child has a strip of 8 squares, and counters. Place 1 counter in any square and see that 1/8 of the whole strip has been covered. Continue placing counters one by one and naming the fractional part covered. Also, turn strip over, fold and cut into halves, fourths, and eighths. Say: 1/2 of 8 = 4, etc.

Figure 8.5

Set	Number in set	Number shaded	Number not shaded	Fraction shaded	Fraction not shaded	Equation
	4	3	1	3/4	1/4	3/4 + 1/4 = 4/4 = 1
	3	2	1	2/3	1/3	2/3 + 1/3 = 3/3 = 1
	5	3	2	3/5	2/5	3/5 + 2/5 = 5/5 = 1
	7	2	5	2/7	5/7	2/7 + 5/7 = 7/7 = 1
	8	5	3	5/8	3/8	5/8 + 3/8 = 8/8 = 1

Consider these ideas:

A fractional number shows a partitioning or division of:

(1) one whole or several wholes; or

(2) a set of wholes

Sometimes, we use both of these ideas for the same basic set: A dozen eggs may be thought of as *one whole dozen*, or as a *set of 12 whole eggs*. We may think of 1/2 of 1 dozen, or 1/2 of 12 things.

DO-IT-YOURSELF

Test the fractional knowledge of your pupils with individual charts that they fill in themselves. Hand out squares and these directions: Draw a set in each square and circle the fractional part of each set to illustrate each fraction in the list (Figure 8.6).

Figure 8.6

1. 1/4
2. 3/8
3. 2/5
4. 5/9
5. 1/10
6. 6/7
7. 2/3
8. 5/10

Fractions — Bits of This and That 173

FRACTIONAL EQUIVALENCE CHART

Figure 8.7

	Halves	Thirds	Fourths	Fifths	Sixths	Sevenths	Eighths	Ninths	Tenths	Elevenths	Twelfths
1/2	1		2		→3		4		5		6
1/3		1			→2			3			4
2/3		2			4			6			8
1/4			1				2				3
3/4			3				6				9
1/5				1					2		
2/5				2					4		
3/5				3					6		
4/5				4					8		
1/6					1						2

$$\tfrac{1}{2} + \tfrac{1}{3} = \tfrac{3}{6} + \tfrac{2}{6} = \tfrac{5}{6}$$

Filling in the chart is an excellent learning activity. Children discover that the denominator is a divisor. If the denominator is 2, there will be a fraction equal to 1/2 in every second space; if 3, the equivalent fractions will be in every third space; and a denominator of 4 divides the row into 4-space units. When adding 1/2 and 1/3, move along the rows to discover that 6 is the least common denominator (Figure 8.7). $\frac{1}{2} + \frac{1}{3} = \frac{3}{6} + \frac{2}{6} = \frac{5}{6}$

FRACTIONS ON THE MULTIPLICATION CHART

The multiplication chart is full of fractional equivalents. To raise a fractional number to higher terms, multiply it by the identity factor, 1, written in some fractional form as 2/2, 3/3, 4/4 and so on.

With two fingers point out the factors 1 and 2. These numerals will be thought of as the numerator and denominator of the fractional number 1/2. Multiply 1/2 by 2/2 by sliding the fingers across the rows 1 and 2 to the 2 column. Both terms of the fractional number are now multiplied by 2. 1/2 x 2/2 = 2/4. Continue to slide the fingers to the 3 column. 1/2 x 3/3 = 3/6; to the 4 column. 1/2 x 4/4 = 4/8; to the 5 column. 1/2 x 5/5 = 5/10; et cetera. When your fingers reach the end of the chart you have not reached the end of the fractions equal to 1/2, for each fractional number belongs to an infinite equivalence class. You could keep right on multiplying by higher and higher fractional numbers equal to 1, and form an unending set of numbers equal to 1/2 (Figure 8.8).

Try sliding the fingers along the rows 3 and 4 to find fractional numbers equal to 3/4. Try 7/8, 8/7, 4/9. Try 6/8 and compare to 3/4.

Figure 8.8

×	1	2	3	4	5	6	7	8	9	10
→1	1	2	3	4	5	6	7	8	9	10
→2	2	4	6	8	10	12	14	16	18	20
3	3	6	9	12	15	18	21	24	27	30
4	4	8	12	16	20	24	28	32	36	40
5	5	10	15	20	25	30	35	40	45	50
6	6	12	18	24	30	36	42	48	54	60
7	7	14	21	28	35	42	49	56	63	70
8	8	16	24	32	40	48	56	64	72	80
9	9	18	27	36	45	54	63	72	81	90
10	10	20	30	40	50	60	70	80	90	100

COMMON DENOMINATORS

1. To add 3/4 and 4/5 (Figure 8.9):

Use the denominators 4 and 5 to obtain the least common denominator, 20. To change 3/4, slide fingers along Rows 3 and 4 to the *5* column. The fingers will be on 15/20. 3/4 x 5/5 = 15/20. To change 4/5 slide fingers down columns 4 and 5 to the *4* Row. The fingers will be on 16/20. 4/5 x 4/4 = 16/20. 15/20 + 16/20 = 31/20

Figure 8.9

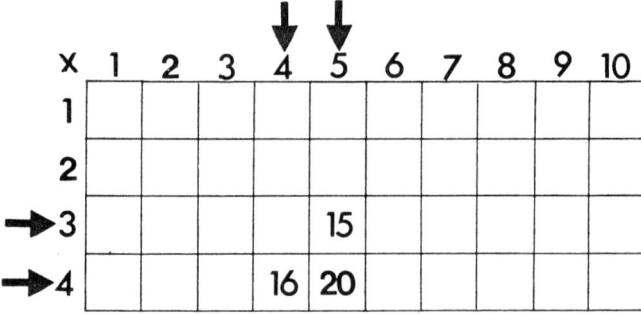

2. To add 3/4 and 4/5 (Figure 8.10):

Cut multiplication chart apart by rows. For 3/4, place rows 3 and 4 together, and to represent 4/5, use rows 4 and 5. The least common denominator is 20. Place these denominators directly under each other. Add the numerators 15 and 16. 15/20 + 16/20 = 31/20

Fractions — Bits of This and That

Figure 8.10

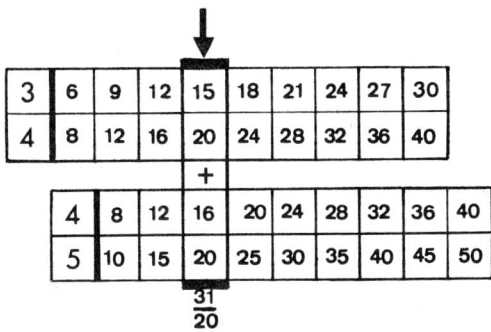

EQUIVALENT FRACTION CRISSCROSS

Frame any four numbers on the multiplication chart and you have a pair of equivalent fractions (Figure 8.11). ad = bc shows $\frac{a}{b} = \frac{c}{d}$.

Figure 8.11

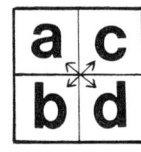 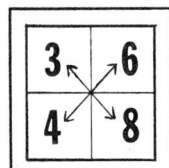

COORDINATE GRAPHS OF EQUIVALENT FRACTIONS (Figure 8.12)

Graphing equivalent fractions on a coordinate grid is a review project for the class, and informally introduces the concept of slope. Have the class participate in a discussion of the results:

1. Graph of 1/2: (2, 1), (4, 2), (6, 3), (8, 4), (10, 5)

2. Graph of 1/3: (3, 1), (6, 2), (9, 3)

3. Graph of 3/4: (4, 3), (8, 6)

4. Other graphs shown: 1/4, 2/3, 1, 5/4, 2, 10.

5. The arrows indicate infinity for fractional equivalence.

6. The proper fractions lie below 1 on the chart.

7. The improper fractions lie above 1 on the chart.

8. Where would the graph for 100 lie?

9. The ordered pairs for the fractional numbers are given as (denominator, numerator) so as to retain the concept of slope, which may be thought of as "rise" over "run."

10. What fractional numbers would graph on the horizontal or x-axis? ($\frac{0}{x}$) The x-axis has a slope of 0.

11. Could you find a fractional number that would graph on the vertical or y-axis? Why not? There is no number that can name the slope of the y-axis. Try to graph 200 (200/1) on the grid. Try 1,000. We see we cannot have ($\frac{y}{0}$). (See Never Divide by Zero, Chapter 6, Multiplication - Division.)

12. The hypotenuse of the right triangle with 45 angles, rises to the right, and has a slope of 1. (See *A Giant Protractor*, Chapter 8.)

Figure 8.12

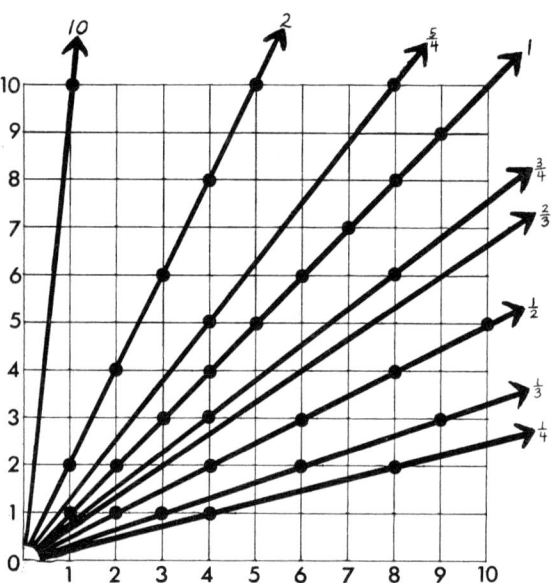

A LINE YOU MUST CROSS – FRACTIONAL EQUATIONS (Figure 8.13)

1. Make a suitable addition equation for each row.

2. Make a suitable subtraction equation for each row. In one-half of the equations use decimal fraction notation. Example: 10 - 9.5 = .5

COLOR-SQUARED SUBTRACTION WITH REGROUPING – WHEN A FRACTION NEEDS A FRIEND

Pupils are to say orally what operation has been represented in Figure 8.14 what has taken place, and the result. After this discussion, they state their ideas of how the problem might be written, renaming shown, and other work performed (Figure 8.14).

Fractions – Bits of This and That

Figure 8.13

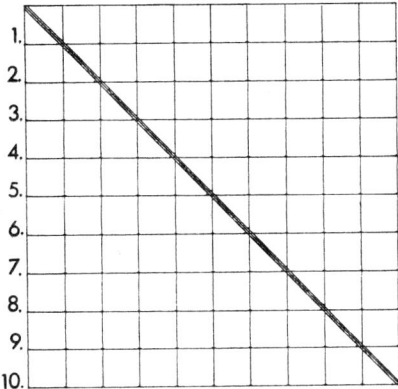

Figure 8.14

$$2\text{-}1/2 = 2\text{-}2/4 = 1\text{-}6/4$$
$$-\phantom{2\text{-}}3/4 = \phantom{2\text{-}}3/4 = \phantom{1\text{-}}3/4$$
$$\phantom{2\text{-}1/2 = 2\text{-}2/4 =\ }1\text{-}3/4$$

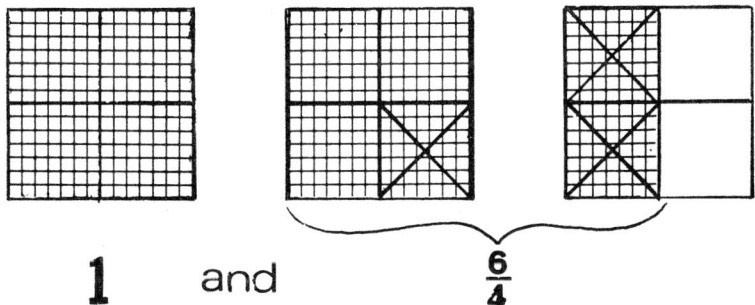

It is possible to perform the subtraction 1/2 - 3/4 if you have a helpful whole number standing by for renaming.

UNITED STATES CURRENCY

The American money system is a decimal system, and the basic unit is the American dollar. Fractional parts of the dollar are named as tenths and hundredths, and in approved money notation, the decimal point is used and these two places are filled in, with zeros if necessary.

Drawing pennies to show each amount provides an interesting learning activity for young children (Figure 8.15).

178 *Fractions – Bits of This and That*

Figure 8.15

Paste pennies on a grid to show 56¢, or 56/100 of a dollar.

FILL A PIG – DIFFERENT WAYS TO BANK YOUR MONEY

Children enjoy filling in squares, and neatness is practiced as they record the many different ways they might have certain amounts of money in their piggy banks.

Dimes, Nickels, Pennies (Figure 8.16)

	d	n	p
	1	1	2
17¢	1		7
		3	2
			17

Quarters, Nickels, Pennies (Base 5 part of the U.S. money system) See Figure 8.17

Extend chart.

Figure 8.17

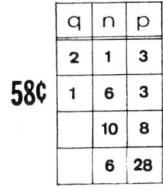

Fill in the chart in Figure 8.18 with the fewest bills and coins that can produce each amount.

Figure 8.18

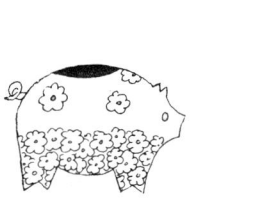

	Dollars	Half-dollars	Quarters	Dimes	Nickels	Pennies
$0.35			1	1		
$0.43			1	1	1	3
$0.88						
$4.83						
$1.99						

Fill in the squares in Figure 8.19 with exactly 6 coins to show the amounts given.

Figure 8.19

$0.18	ⓝ	ⓝ	ⓝ	ⓟ	ⓟ	ⓟ
$0.31	ⓓ	ⓝ	ⓝ	ⓝ	ⓝ	ⓟ
$0.42	ⓠ	ⓝ	ⓝ	ⓝ	ⓟ	ⓟ
$1.00	ⓠ	ⓠ	ⓠ	ⓓ	ⓓ	ⓝ

DECIMAL PLACE VALUE – WHOLE NUMBERS AND FRACTIONS

Sets of squares A and B are used to picture decimal number concept. Have pupils explain each set by pointing out the unit in each case and the relation of the other squares, or parts of squares, to the unit (Figure 8.20).

PERCENTS AND THE HUNDRED SQUARE – GRAPHIC MODELS OF PERCENTAGE

Directions to pupils: Per cent means "to a hundred." To help you understand this ratio, use a 100-square grid. Touch 1 square. You have touched

1% of the hundred squares. Run your finger along 10 of the squares. You have touched 10% of the squares. If you touch 15 squares, what percent of the 100 squares did you touch? Touch 1/2 of the squares. What percent did you touch? Explain.

Use a set of ten squares. Touch 1/2 of this set. Is this 50%? Can you explain this as 5/10 = 50/100, or .5 = .50? What is 50% of 8 squares? of 1000? of 5? See Figure 8.21

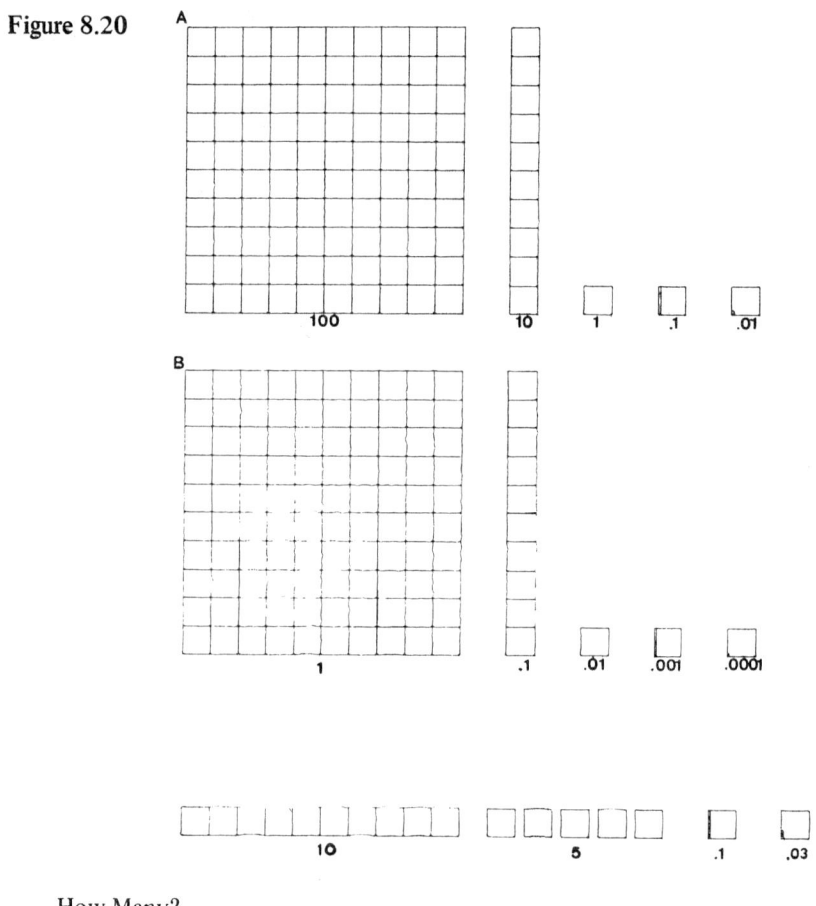

Figure 8.20

How Many?
__1__ ten __5__ ones __1__ tenth __3__ hundredths = 15.13, or 15 13/100.

Fractions – Bits of This and That 181

Figure 8.21

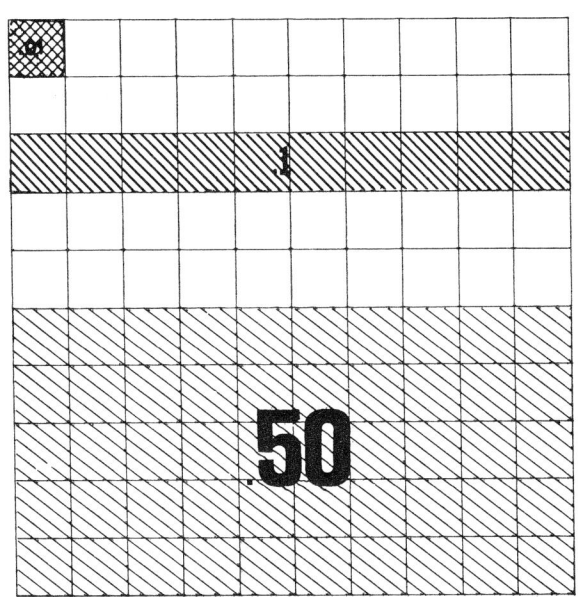

100	100%	1.00	$\frac{100}{100}$	$\frac{1}{1}$	1
50	50%	.50	$\frac{50}{100}$	$\frac{1}{2}$	.5
10	10%	.10	$\frac{10}{100}$	$\frac{1}{10}$	.1
1	1%	.01	$\frac{1}{100}$	$\frac{1}{100}$	.01

Draw blue dots on 25 of the squares. Study this notation: 25/100 = .25 = 25%. 25% means a ratio of 25 to 100, or 25:100. Do you see that 1/4 of the large square is dotted? Consequently, 1/4 = 1:4 = 25/100 = .25 = 25%. Your chart is also a graphic model for 3/4. How? Write this in all the ways set forth in the 1/4 example above (Figure 8.22).

Do you sometimes confuse 7.5% with 75%: 7.5% means 7-1/2%, and featured on a chart would show only 7-1/2 colored squares (Figure 8.23).

Figure 8.22

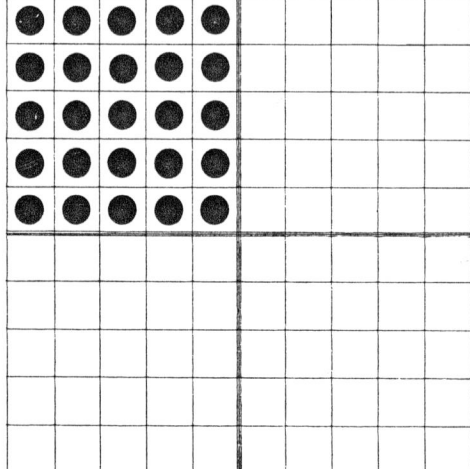

Figure 8.23

.01	.20	.28	.225	.015	.08	.2	1.30	.75	.075	Decimal Fraction
$\frac{1}{100}$	$\frac{20}{100}$	$\frac{28}{100}$	$\frac{225}{1000}$	$\frac{15}{1000}$	$\frac{8}{100}$	$\frac{2}{10}$	$\frac{130}{100}$	$\frac{75}{100}$	$\frac{75}{1000}$	Common Fraction
1%	20%	28%	22.5%	1.5%	8%	20%	130%	75%	$7\frac{1}{2}$%	Percent

PAVE A ROAD – USE STORY PROBLEMS AND NUMBER LINES FOR FRACTIONAL OPERATIONS (Figure 8.24)

Line A: The paving crew paves .2, .7, and .9 miles on three successive days. How many miles did they pave altogether? .2 + .7 + .9 = 1.8 mi.

Line B: The paving crew averages .4 of a mile of new pavement for 5 days. How many miles did this crew pave altogether? 5 x .4 = 2 mi.

Line C: The crew intended to pave .8 of a mile per day for 5 days, but they had only worked 1/2, or .5, of a day when it began to rain. How much of the road was paved? .5 x .8 = .4 mi.

Practice subtraction: The crew intends to pave 2.3 miles in two days. They paved 1.5 miles on the first day. How many miles are left to be done on the second day? 2.3 - 1.5 = .8 mi.

Practice division: The crew intends to pave 6 miles. If they can pave at the rate of .5 mile per day, how many days will it take them to complete the job? 6 ÷ .5 = 12 days

Fractions – Bits of This and That   183

Figure 8.24

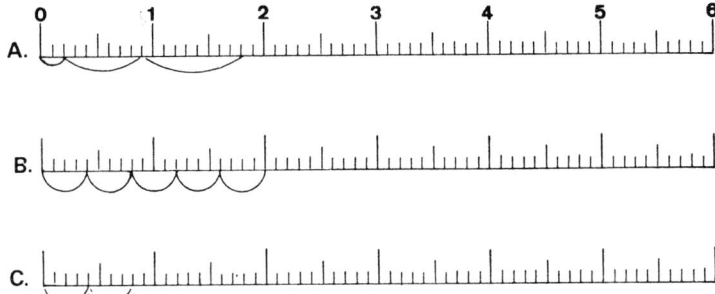

PAVE A ROAD – USE OF NUMBER LINE
FOR ALL FRACTIONAL PROBLEMS (Figure 8.25)

A. 5 days, 1/2 mi. per day
B. 5 days, 3/4 mi. per day
C. 1/2 day, 3/4 mi. per day

Figure 8.25

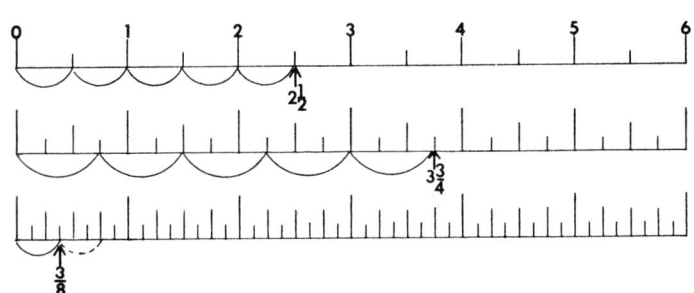

FRACTIONAL AREA (Figures 8.26, 8.27)

.8 x .4 = .32 is similar to an 8 x 4 array.

Figure 8.26   Figure 8.27

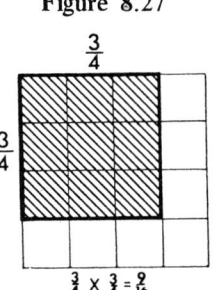

$\frac{3}{4} \times \frac{3}{4} = \frac{9}{16}$

MULTIPLICATION CHARTS FOR FRACTIONS

Compare with the whole number chart (Figure 8.28).

Figure 8.28

×	1	2	3	4
1	1	2	3	4
2	2	4	6	8
3	3	6	9	12
4	4	8	12	16

×	1/4	2/4	3/4	4/4
1/4	1/16	2/16	3/16	4/16
2/4	2/16	4/16	6/16	8/16
3/4	3/16	6/16	9/16	12/16
4/4	4/16	8/16	12/16	16/16

×	.1	.2	.3	.4
1	.1	.2	.3	.4
2	.2	.4	.6	.8
3	.3	.6	.9	1.2
4	.4	.8	1.2	1.6

×	.1	.2	.3	.4
.1	.01	.02	.03	.04
.2	.02	.04	.06	.08
.3	.03	.06	.08	.12
.4	.04	.08	.12	.16

A FRACTION OF A FRACTION

Multiplication of fractions may be shown as repeated addition: 1/4 + 1/4 + 1/4 or 3 × 1/4. One-half of 3/4 is equal to one-half of 1/4 added three times (Figure 8.29).

Figure 8.29

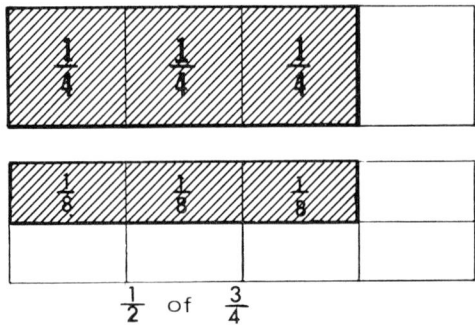

1/2 of 3/4

RECIPROCALS – CAN YOU PICTURE THEM?

A reciprocal is the multiplicative inverse of a fraction. Multiplying a fraction by its reciprocal yields 1 as a product. For example: 3/2 is the reciprocal of 2/3 (Figure 8.30).

Fractions – Bits of This and That

Figure 8.30

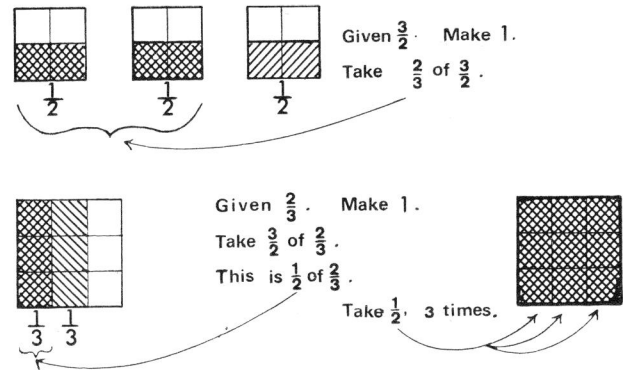

DISTRIBUTIVE LAW OF MULTIPLICATION APPLIED TO FRACTIONAL NUMBERS

Fold the apple array along the black line to show 2 x 3 and 2 x 1/2.

Figure 8.31

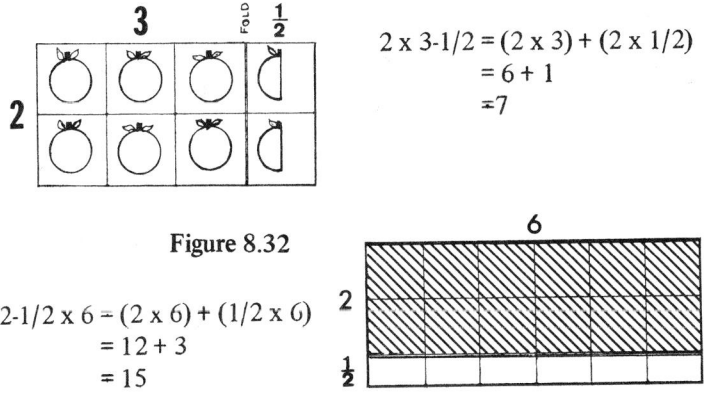

2 x 3-1/2 = (2 x 3) + (2 x 1/2)
= 6 + 1
= 7

Figure 8.32

2-1/2 x 6 = (2 x 6) + (1/2 x 6)
= 12 + 3
= 15

Figure 8.33   2-1/2 x 3-1/2 = (2 + 1/2) x (3 + 1/2)
= (2 x 3) + (2 x 1/2) + (1/2 x 3) + (1/2 x 1/2)
= 6 + 1 + 1-1/2 + 1/4
= 8-3/4

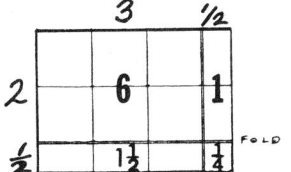

Use two different ways to obtain the products for Card III (Figure 8.34).

Figure 8.34

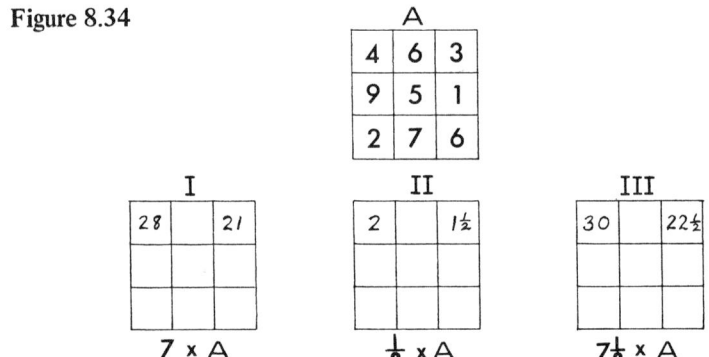

COMPARISON PUZZLES (Figure 8.35)

Figure 8.35

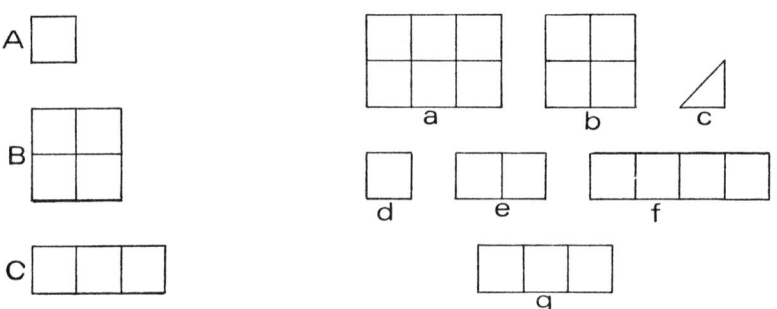

What is the measure of each of the regions above:

1. If A is one whole unit?

   Answers: a. 6; b. 4; c. 1/2; d. 1; e. 2; f. 4; g. 3

2. If B is one whole unit?

   Answers: a. 1-1/2; b. 1; c. 1/8; d. 1/4; e. 1/2; f. 1; g. 3/4

3. If C is one whole unit?

   Answers: a. 2; b. 1-1/3; c. 1/6; d. 1/3; e. 2/3; f. 1-1/3; g. 1

Fractions — Bits of This and That 187

GREAT DISCOVERY — UNDERSTANDING EUCLID'S ALGORITHM FOR FINDING G.C.D.

When renaming a fraction in lowest terms, it is helpful to find the greatest common divisor of the numerator and denominator. The Counting and Multiplication Charts help pupils discover various methods of arriving at the G.C.D. One very interesting method is Euclid's Algorithm.

Example: What are the common divisors of 6 and 9? What is the G.C.D.? Have a class examine both Charts and discuss divisors of these numbers.

1. All numbers equal to or less than 9 are potential divisors of 9.

2. 6 is less than 9, so a common divisor of both must be 6 or less than 6.

3. The difference between 9 and 6 is 3. A common divisor of 9 and 6 must also be a divisor of that difference, 3, or there would be a remainder of 3. In other words, the G.C.D. must be 3 or some divisor of 3. The only divisors of 3 are 3 and 1, and these are also divisors of both 6 and 9.

4. 3, the difference between 9 and 6, is the G.C.D. of these numbers. (This was an exciting discovery when divisors were explored on a counting chart. More discoveries disclose what Euclid aready knew.)

Try another example: What is the G.C.D. of 18 and 24? Study the charts as before.

(1) The difference between 18 and 24 is 6. 6 is a divisor of both 18 and 24, and no greater one can be found. (The Multiplication Chart also shows that the divisors of this difference, 6, are 3, 2, and 1, and that these are divisors of 18 and 24.)

One more try: What is the G.C.D. of 9 and 15?

(1) Their difference is 6. This time, the "first difference" exactly divides neither 9 nor 15. By the Multiplication Chart, 3 is a common divisor of both 9 and 15. 3 is the difference between 6 and 9. Here a "second difference" was needed.

Last try: What is the G.C.D. of 14 and 24?

(1) The "first difference," 10, is a divisor of neither 14 nor 24. The "second difference," 14 - 10, or 4, is not a divisor of either 14 or 24. The "third difference" is 10 - 4, or 6. The rule of "differences" is not working. The Multiplication Chart shows 2 as the G.C.D. 10 - 8, or 10 minus two 4's, yields the correct answer, 2.

Try it Euclid's way:

$$14\overline{)24}$$
$$\underline{-14}\quad 1$$
1st difference → $10\overline{)14}$
$$\underline{-10}\quad 2$$
2nd difference → $4\overline{)10}$
$$\underline{-8}\quad 2$$
Last difference → $2\overline{)4}$
is the G.C.D.
$$\underline{\ 4}$$
$$.\ 0$$

BE DIFFERENT WHEN YOU RENAME ONE-HALF

Children are asked to stretch their imaginations and be original as they help fill the 1/2 chart shown in Figure 8.36.

Figure 8.36

$\frac{60}{120}$	$\frac{99}{198}$	$\frac{75}{150}$	.500	$\frac{50}{100}$
$\frac{33}{66}$	$4 \div 8$	$\frac{1}{3} \times 6$	$\frac{45}{90}$	$\frac{1}{8} + \frac{3}{8}$
$\frac{3}{4} - \frac{1}{4}$	$\frac{3}{8} \cdot \frac{3}{4}$	$\frac{1}{2}$	$1 \div 2$	$\frac{1}{4} + \frac{1}{4}$
$\frac{8}{16}$	$1 - .5$	$5 \times .1$	$\frac{4}{8}$	$\frac{1}{10}$ of 5
$\frac{.015}{.030}$	$16\overline{)8}$	.5	$\frac{9}{18}$	$\frac{.06}{.12}$

CRAZY FRACTION MULTIPLICATION CHART (Figure 8.37)

All products in this crazy chart must be written in fraction form. Let teams devise these Crazy Charts. Have a contest and a crazy prize for the "craziest."

Figure 8.37

×	1	2	3	4	5
1	$\frac{3}{3}$	$\frac{8}{4}$	$\frac{18}{6}$	$\frac{36}{9}$	$\frac{50}{10}$
2	$\frac{1}{3} + \frac{1}{6}$	$\frac{8}{2}$	$\frac{18}{3}$	$\frac{48}{6}$	$4 \times 2\frac{1}{2}$
3	$5 \times \frac{3}{5}$	$\frac{1}{2}$ of 12	$\frac{9}{1}$	$\frac{36}{3}$	$2 \times 7\frac{1}{2}$
4	$7 \times \frac{4}{7}$	$\frac{24}{3}$	$\frac{60}{5}$	$18 - \frac{4}{2}$	$\frac{24}{2} + 8$
5	$\frac{25}{5}$	$\frac{81}{9} + 1$	$\frac{5}{3} \times 9$	$18 + \frac{4}{2}$	$\frac{100}{4}$

A WHALE ON A SCALE – THINK BIG

From the answers given in Figure 8.38 pick a sensible weight for a full-grown whale.

USABLE BULLETIN BOARD

Put up a picture of a large whale. Cut apart large squares, one for each suggested numeral and one for the decimal point. Each day allow a different child to arrange the squares to form his idea of the whale's weight. (Of the answers suggested here, the first one listed is the best answer.)

Figure 8.38

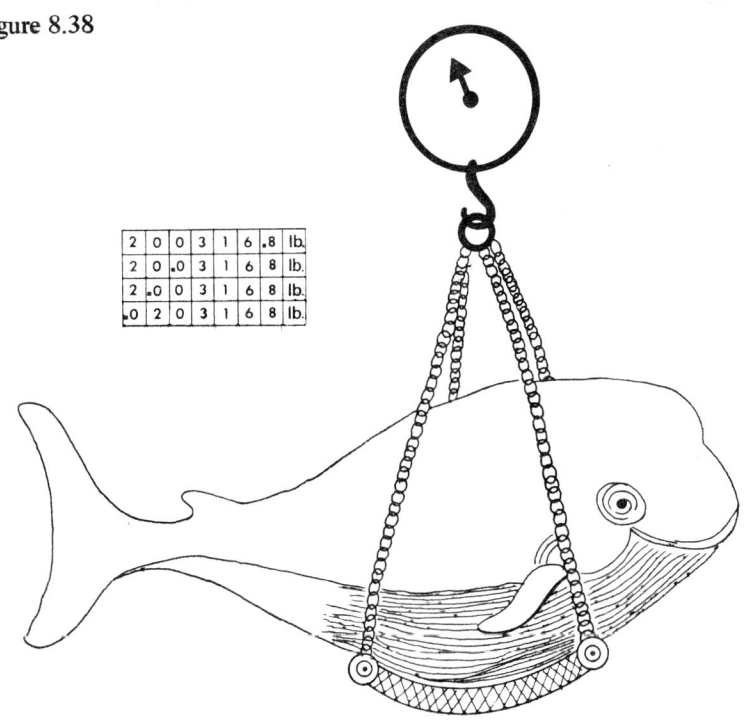

A SNAIL ON A SCALE – THINK SMALL

Now try your luck with this midget, the snail (Figure 8.39).

Figure 8.39

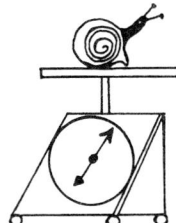

9.

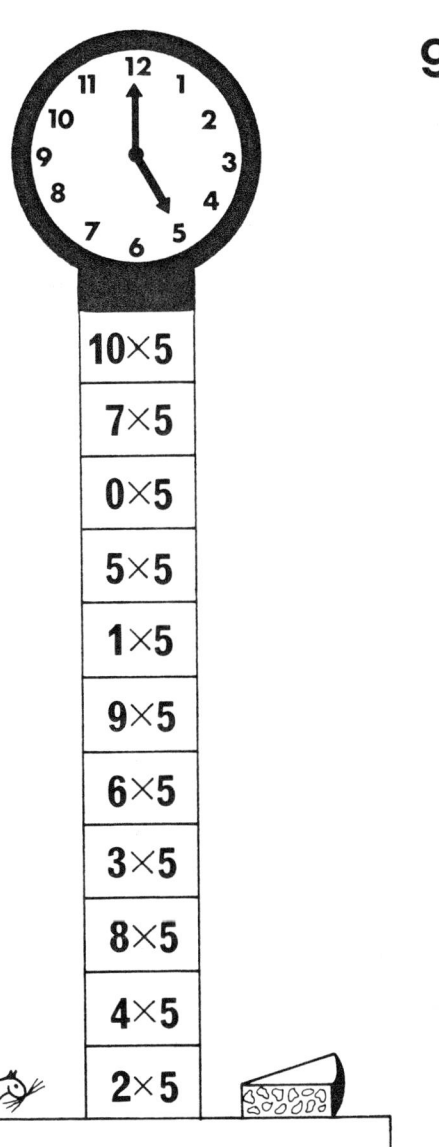

9

 **It's Your Turn - Games**

— the boys usually beat the girls — playing baseball

Included in this chapter are simple games and enrichment activities, some suitable for a child to play alone, and others designed to be played with a classmate. Many may be adapted for team use and enjoyed by the entire class.

Although many exercises containing pleasure aspects are scattered throughout the book, some games, with modifications, are applicable to all the whole number operations, and also may be conveniently used for fractional and integral purposes. It was felt that these should not be included in any chapter, but collected and presented here. Just as a top-notch skater continues to practice to put shine on his performance, the student needs drill to polish up and keep his skills bright.

A relaxed, comfortable attitude towards a subject is a second reason for including games in the lessons. Analyzing game situations, manipulating counters, and meeting numbers and relationships in new situations, provide the challenge, interest, and fun that help establish a positive viewpoint towards mathematics.

As a teacher, you are always searching for new games. You can be an inventor of a fresh game by giving a new twist to the old — just use squared paper and give your imagination free rein. Be sure to

It's Your Turn – Games

revise the games to suit your class and circumstances. Set up simple rules for young children and more complicated ones for the older grades. You're the umpire!

For maximum use, laminate all charts. Whenever possible, use counters instead of writing on the charts. When written answers are necessary, place small pieces of masking tape over each previously written answer. Children write on the tape, then pull it off to check their answers.

Let squared paper help you when you are planning signs and materials for a bulletin board.

Have fun!

Figure 9.1

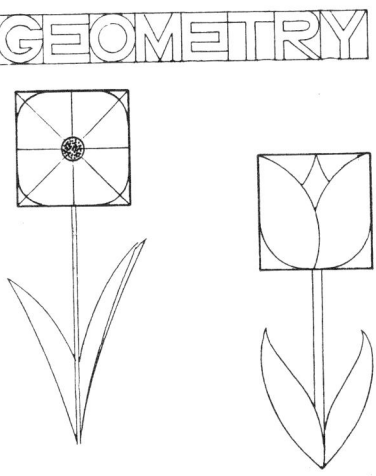

POISON SQUARE (Figure 9.2)

Two players take turns – rolling 1 die and moving the designated number of squares – attempting to be the first to reach the goal – but must return to GO every time they land on a POISON SQUARE.

First they decide on which squares will be poison. Possible POISON SQUARES:

1. All numbers that end in 5.
2. All multiples of 9.
3. All primes.
4. All multiples of 6, plus 1.

Figure 9.2

GO	1	2	3	4	5	6	7	8	9
19	18	17	16	15	14	13	12	11	10
20	21	22	23	24	25	26	27	28	29
39	38	37	36	35	34	33	32	31	30
40	41	42	43	44	45	46	47	48	49
59	58	57	56	55	54	53	52	51	50
60	61	62	63	64	65	66	67	68	69
79	78	77	76	75	74	73	72	71	70
80	81	82	83	84	85	86	87	88	89
★	98	97	96	95	94	93	92	91	90

WOLF AND LITTLE PIG (Figure 9.3)

1. Play with a friend. Write problems on windows, steps, and door.

2. Take turns.

3. Only one problem may be answered at a turn.

4. The Pig starts at the top of the house and each time he gives the correct answer, cover the window with a counter.

5. The Wolf comes a step nearer every time he answers correctly.

6. Both try to answer the door problem first. Piggy's answer closes the door and keeps out Mr. Wolf. Wolf's answer gets him inside!!!

It's Your Turn – Games

Variation:

Team game: Taking turns, the Little Pig team tries to close every window and finally the door before the Wolf team gets up the steps and into the house.

For a reusable game do not write the problems on the game board. Selected flashcards are shuffled, and drawn in turn. For each correct answer, a counter is placed on a window, a step, or the winning door to indicate the progress. Or, problems, written on separate squares, are placed on the board and turned over when correctly answered.

GO

Two players take turns rolling a die, and moving their counters the indicated moves along the green line. Landing on a yellow square means a loss of one turn, and landing on a large dot means to return to GO and begin again. Creative children will want to make similar games on larger grids. Either a spinner, numbered cards to be drawn or a numbered foam rubber cube may be used instead of a die (Figure 9.4).

Figure 9.4

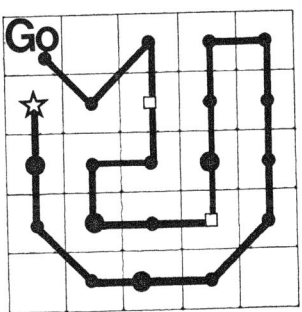

THE BEAR WENT OVER THE MOUNTAIN

This honey-loving animal tries to get to the other side of the mountain (Figure 9.5). Children use a cardboard bear, fingers, or a pointer to act out each trip. Invent fresh, new games by changing the titles and pictures. Use the same difficult facts and try: Jack and Jill Went Up the Hill, Will They Tumble Down? Climb the Tree to Pick Sweet Cherries, Climb Back Down to Eat Them.

Figure 9.5

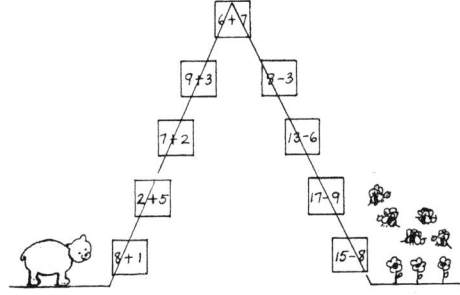

LUCKY TENS

Add two or more adjacent numbers to make a sum of ten. Move across the rows, down the columns, or diagonally. Several samples are shown in figure 9.6. (So many "tens" may be found that three boards are suggested. Use one for rows, one for columns, and one for diagonals. Also it helps to write down each problem as it is found. Young children use only the horizontal form.)

Figure 9.6

3	7	8	1	9	5	5	8	3	4
1	3	4	2	8	3	6	4	5	7
9	5	4	2	4	7	3	6	1	3
6	9	8	6	1	7	2	3	9	1
4	2	2	2	1	5	7	5	5	3
7	6	8	1	4	3	1	2	3	6
3	2	8	5	7	4	9	3	2	5
5	8	1	2	4	1	9	5	4	1
8	2	3	2	6	4	4	7	3	4
7	5	5	3	3	1	3	6	1	2

RULE PICTURES

The rules tell what dots to connect to discover the secret pictures.
Rule: Multiples of 7 in order (Figure 9.7).
Rule: Sums of 13 (Figure 9.8).

Figure 9.7   Figure 9.8

 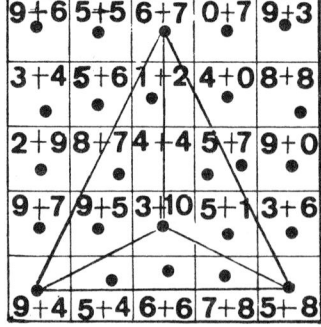

It's Your Turn – Games 197

GO AROUND THE BLOCK OR ON A LONG SPACE TRIP

Frames of squares resemble a city block. Each set of consecutive numerals makes up one problem. The player adds around to see if he can get back home. With squares and children's drawings, the teacher can make use of this same idea to create many enticing practice games (Figure 9.9). CAFETERIA – from soup to pie; THROUGH THE ZOO – see monkeys, elephants, and the gnu; JUNGLE TRIP – rubber trees and orchids; SAFARI – lions and zebras; AROUND YOUR TOWN – bank, firehouse, school and church; AROUND THE WORLD – igloos in Alaska, Heidi's Alps in Switzerland, and junks in Hong Kong; UNDER THE SEA – a shark, a whale and a sunken ship. A space trip is a must!

Variation: Tape a game to the chalkboard, write a number in the center, and have a pupil add, or multiply, the center number with each of the trip numbers. This may be an oral exercise, or the answer to each problem may be written directly outside it.

Figure 9.9

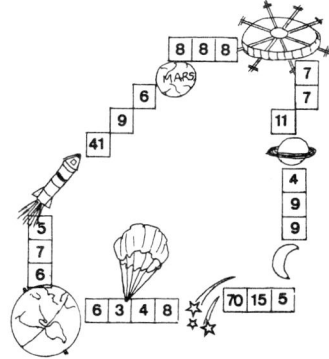

JUMPER, THE CRICKET (Figure 9.10)

The Three Jump: Add. Jumper will prove you are right – or wrong.
Get Set: SETS tells Jumper how many jumps to make.

ZIP-O'S

To play Add-0, each pupil has a card of sums. As the teacher shows one flash card at a time, the pupils add mentally and if the sum is on his card he covers it with a counter. Traditionally, covering a line in any direction is an Add-0, but other rules for winning may be adopted. Mult-0 is a similar game for multiplication. To play Fact-0, numbers from 0 to 12 are drawn by the caller and problems are covered by the pupils. For example, 7 is called, and problems 6 + 1 and 8 - 1 are covered (Figure 9.11).

Figure 9.10

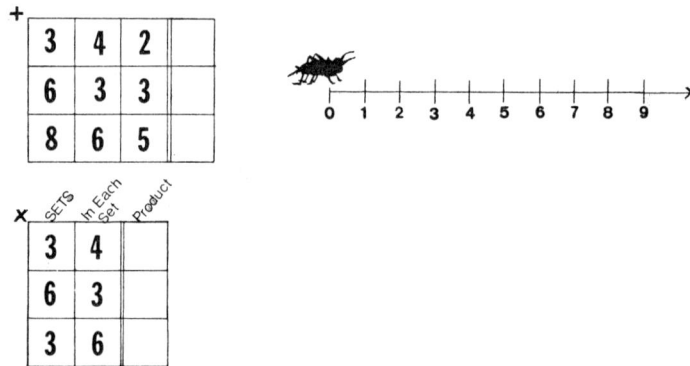

Figure 9.11

Add-O

8	5	0	4	18
14	2	12	3	9
1	9	A	6	5
7	11	19	16	8
10	15	13	20	17

Mult-O

45	16	25	40	4
32	3	81	20	10
12	36	M	21	8
18	72	9	24	5
6	30	63	27	15

Fact-O

8+2	5−5	6+1	2×2	4+5
16−6	7+5	3×3	4+4	8−3
3×4	5+6	F	9−7	15−8
5×0	18−9	2×3	10−3	6+4
5+0	8+3	2+3	12−4	13−6

To play Add-0-Win, Productivity and Fraction-Action, two players alternate selecting and filling in one square at a time. Player filling most squares wins the game. For team play, place a large 5 x 5 grid on the magnetic board. Girls place red counters whenever they give a correct sum, and boys place blue counters. Turns are taken. When all the squares are filled, the team that placed the greatest number of counters wins (Figure 9.12).

Figure 9.12

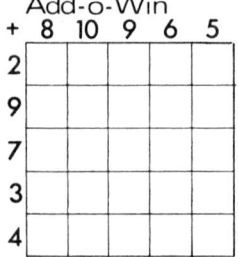
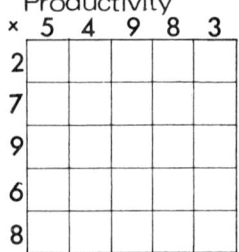
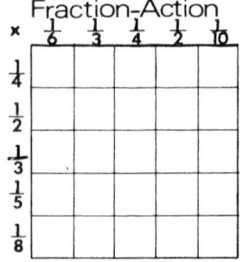

It's Your Turn – Games   199

FOOTBALL

The field is easily drawn on squared paper, and the 10-yard lines marked. A small magnetic football is placed on the 50-yard line. Opposing teams toss for the first go. The teacher shuffles a selected set of flashcards, and shows the first card to the first member of the first-go team. If the answer is correct, the ball is moved 10 yards. The other team has a chance at the next card. A correct answer brings the ball back 10 yards, while an incorrect answer gets no move. Each team is trying to reach its own end-zone for a score of 6 points. When this happens, the scoring team immediately gets another card to try for the extra point after touchdown. A time for the final whistle may be set, or the game may end when everyone has had one turn (Figure 9.13).

Figure 9.13

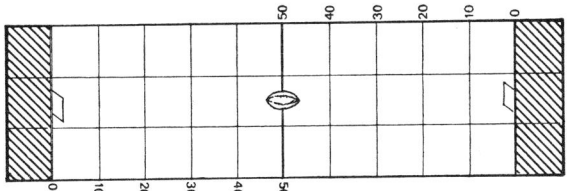

BASEBALL

One person on each team (girls versus boys) is chosen to be the pitcher, and will throw a different ball, (change the large center square), for each opposing batter. Each batter starts at first base and must correctly complete the circuit in order to score a run. Missing at any base is an out, an x is scored, and the teams change sides. Three outs for either side means the game is over. The teacher is the umpire, and may change the bases at any time. As multiplying improves and it becomes difficult to retire the side, the bases may be 2- and 3-digit numbers, as 60, 81, 701, 222, and so forth (Figure 9.14).

Smaller children play a simpler version of the game. Addition is the operation; one of the girls is up, then one of the boys, and so on, until everyone has had a chance, or there have been 3 outs. The teacher does all the changing of the bases and pitched balls, regulating the difficulty of the play to suit individual abilities. (Answers, Figure 9.14 – 11, 13, 5, 14.)

ROLL OVERS

Crease along the lines between the squares, and roll this practice folder over and over in the same direction as you write in the 7th table. To use, say 7 x 1 = 7, fold over and see the answer 7, which lets you know that you were correct. Fold over again, see 2, say 7 x 2 = 14, fold over and you see 14. Continue and you will soon arrive at 7 x 10 = 70. Now your Roll-Over is rolled up and you are back to 1. As you go through the 7th table again, your Roll-Over will unroll. Try it! (On the strip, 6 is behind 42.) Figure 9.15.

Figure 9.14

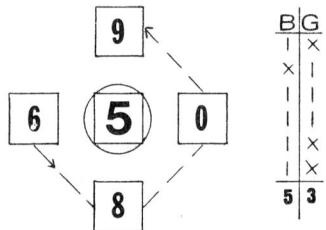

Figure 9.15

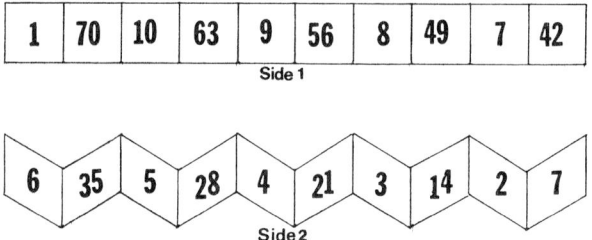

SQUARE NIMS

This game is played on 20 squares. Two children take turns placing counters to cover the playing strip. One, two, three, or four counters may be placed at each turn. The squares must be played in order from the first to the last, and the person who can fill in the last square wins (Figure 9.16). Can you work out a winning system? Variations: The player who must fill in the last square loses. Is there a winning system now? Try playing the game on a 15-strip with a 1, 2, or 3 play at a turn, and figure out a good winning method.

Figure 9.16

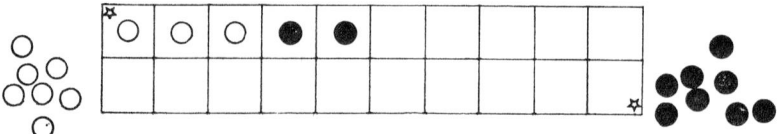

It's Your Turn — Games 201

BATTLESHIP

Two children play. Each player marks off a 10 x 10 grid with x and y axes, and numbers the points along each axis from +5 to -5. Players sit back to back, and each places counters on five different points on his own coordinate board, to indicate the locations of his five battleships. Taking alternate turns, each player calls out an ordered pair of numbers. If the enemy has a battle ship at the called point, he must declare the ship sunk and remove this counter from his coordinate board (Figure 9.17). The player who sinks all his enemy's ships before he loses his entire fleet, is the winner. He is said to be an admirable player, and is dubbed "Admiral" until he loses a game and his title. (Only one player's grid is shown.)

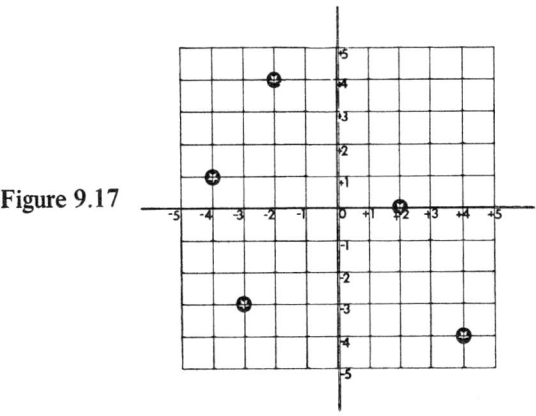

Figure 9.17

PARK THE CARS

Color cars on large squares, one for each product on the multiplication chart. A player gets a point for each car he parks correctly. Car 30 is parked in the 6th row on the 5th floor (Figure 9.18).

Figure 9.18

	5	4	3	2
6	30			
7				
8				
9				

MATH WORDS

Go one block at a time in any direction and use the letters over and over as you search for the math words listed. Can you find others? (Figure 9.19).

1. equal
2. sum
3. angle
4. multiply
5. minus
6. time
7. times
8. square
9. area
10. plus
11. decimal
12. divide
13. seven
14. nine
15. prime
16. semi-circle
17. circle
18. multiple
19. triangle
20. Pi
21. altitude
22. less
23. add
24. cent
25. pair
26. plane
27. even
28. line

Figure 9.19

s	d	a	q	s	e
e	c	e	u	g	l
i	m	n	r	a	n
r	a	i	t	i	v
c	p	l	u	d	e
e	l	y	m	s	n

SIGN GAME (Figure 9.20A)

Give each numeral a positive (+), or negative (-) sign. Add rows and columns. You must end with +8. Think!

Figure 9.20A

1	+	9	+	2	=	6
+				+		+
9	+	2	+	6	=	5
+				+		+
5	+	3	+	1	=	7
=		=		=		=
5	+	10	+	3	=	+8

FOLLOW-ME-GAME (Figure 9.20B)

Pull a number from the hat. Write it down and follow me. Do it right. You'll end up with 3.

A. Write the number.

B. Multiply by 2.

C. Add 6.

D. Divide by 2.

E. Subtract the number you wrote first.

F. Now you have 3.

Figure 9.20B

A	B	C	D	E
12	24	30	15	3
8				3
-6				3
-50				3

MOUSE RAN UP THE CLOCK

This clock is set for the third table (or for adding 3). The child goes up slowly saying each fact and product, as 3 x 10 = 30, but runs down quickly saying just the product. A correct third table earns the chance to say the strikes of the clock, "Bong, bong, bong" when reaching the top (Figure 9.21).

Real fun for everybody: (1) Move a toy mouse up and down; (2) Use a cake pan face and strike it with a ruler.

HOPSCOTCH MULTIPLICATION

Use large squares marked off as shown in Figure 9.22. The court is placed on a flat table. Play regular hop-scotch rules, tossing kernels of corn for pucks instead of the traditional stones. A multiplier is decided and after the puck is tossed, the player multiplies back through all the squares. When a player misses a toss or misses a fact he loses his turn. The player who first successfully moves his puck through all the squares wins a point, and another multiplier is chosen for the next game.

Figure 9.21   Figure 9.22

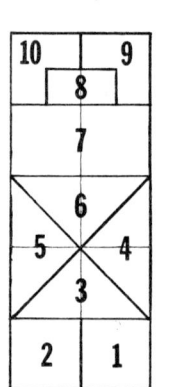

FERRIS WHEEL (Figure 9.23)

Oral Game: Draw a ticket, insert it in the hub of the wheel, start at the bottom, say the answers counterclockwise all the way around and win a point. Your opponent draws a ticket and tries to win a point. After all the tickets have been drawn and used, the game is over. Most points wins.

Written Game: Two wheels taped to chalkboard, two players write answers around the outside as quickly as possible to determine the winner.

Figure 9.23

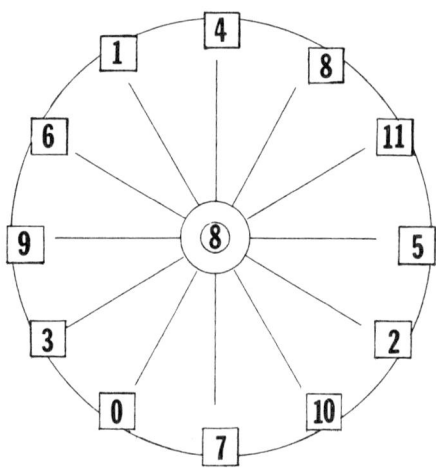

RACE TRACK

Use a small pile of flashcards. The small numbers, 1, 2, and 3, indicate the difficulty of the cards and the number of spaces a player moves when he

It's Your Turn – Games

answers a flashcard fact correctly (Figure 9.24). Let 7 be wild! 7 times any number allows a move of 5 spaces, and 7 x 7 means a move of 10 spaces.

Figure 9.24

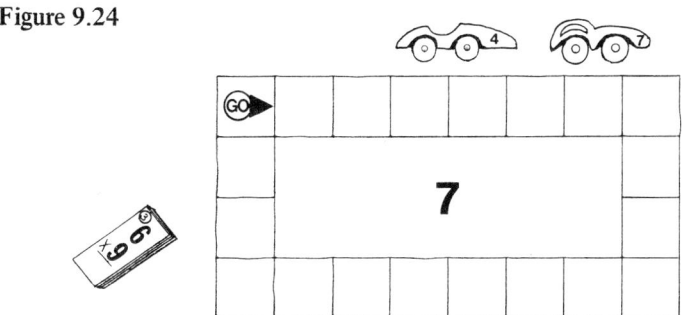

FRAME GAME (9.25)

Here is a game that not only reinforces operation facts, but at the same time emphasizes the matching of geometric forms, a perceptive skill essential to recognizing the letters in reading and the number symbols in mathematics.

Each numeral has a unique frame which the children use to discover the numbers needed to solve the problems. *Example:* ⌐ has a ceiling and a right wall and is the frame for 4. ⌐ has a floor and a right wall and is the frame for 5. ⌐ + ⌐ = 9, or 4 + 5 = 9. Answers: 1) *7*; 2) *4*; 3) *4*; 4) *42*; 5) *11*.

Figure 9.25

5	9	6
3	7	8
4	1	2

1. ⌐ + ⌐ =
2. ⌊ ÷ ⌈ =
3. ⌊⌋ − ⌋ =
4. ☐ × ⌊ =
5. ⌈ + ⌊ + ⌐ =

MAGIC SQUARES (Figures 9.26, 9.27)

Each row, column, and diagonal should add to the same sum.
Hint:
In Magic Square 1, 15 is the sum, and 5 is suggested for the middle square.
In Magic Square 2, 18 is the sum, and 6 is suggested for the middle square.
In Magic Square 3, one number must be changed. Which number is wrong, and what should it be?

If all the sums equal 21, what is a good number for the center square? Is the number 3 a help in filling in these magic squares?

One diagonal in each magic square has a particular pattern that again suggests 3 to be a special number. Explain.

Figure 9.26

1.
4	3	8
9	5	1
2	7	6

2.
8	7	3
1	6	11
9	5	4

Figure 9.27

3.
7	6	11
12	8	3
5	10	9

4.
$\frac{5}{2}$	$\frac{6}{2}$	$\frac{10}{2}$
$\frac{12}{2}$	$\frac{7}{2}$	$\frac{2}{2}$
$\frac{4}{2}$	$\frac{8}{2}$	$\frac{9}{2}$

TIC-TAC-TOE

Two pupils play regular Tic-Tac-Toe rules, except that a player must correctly give the answer to a problem before making his mark in the square. Cover squares with colored counters (Cats and Dogs) so game may be reused. Children may fill in blank grids to make their own games. Any operation or combination of operations may be used (Figure 9.28).

Figure 9.28

8+8	6+7	9+8
7+4	9+9	6+7
4+3	7+9	8+7

6×7	9×4	4×3
3×8	8×9	7×7
9×0	6×3	5×9

35÷7	77÷7	48÷6
81÷9	54÷6	18÷9
63÷9	49÷7	36÷9

PEBBLE TOSS

Two players take turns tossing two pebbles from a small can onto the grid. Each player multiplies the two numbers indicated by his toss, and takes the product for his score. After five spills each, the scores are totaled to see who wins. Variation: Older children toss three pebbles. Two of the pebbles indicate a

2-digit factor and the third pebble becomes the other factor. A player must choose the way to play his digits as he tries to produce the highest score. If the pebbles fall on 4, 6, and 9, he has a choice of 6 x 49, 6 x 94, 9 x 46, 9 x 64, 4 x 96, and 4 x 69 (Figure 9.29). A similar game provides addition practice.

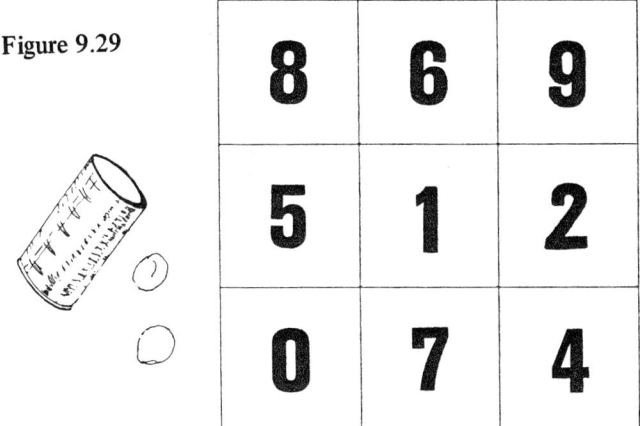

Figure 9.29

SPILL THE BEANS

Players take turns spilling five beans or peas onto a counting chart. Each player records and adds his five addends. The highest sum for each round wins a point. Five points wins a game. Two or more children may play (Figure 9.30).

Figure 9.30

1	2	3	4	5	6	7	8	9	10
11	12	13	14	15	16	17	18	19	20
21	22	23	24	25	26	27	28	29	30
31	32	33	34	35	36	37	38	39	40
41	42	43	44	45	46	47	48	49	50
51	52	53	54	55	56	57	58	59	60
61	62	63	64	65	66	67	68	69	70
71	72	73	74	75	76	77	78	79	80
81	82	83	84	85	86	87	88	89	90
91	92	93	94	95	96	97	98	99	100

PLANT THE CORN

Each player has a 3 x 3 square. He chooses one number from 2 to 12 to write in each of the small squares on his card. Each player has nine kernels of corn to plant in these nine squares. Two dice, or numbered wooden cubes, are tossed, the numbers are added, and if the total is on the player's card, he may plant a kernel of corn in that square. The player who plants all his corn first wins (Figure 9.31). *Variation:* Players must fill their cards in order from left to right along the rows. If 12 is in the first square, then the player must roll 6 and 6, and plant this 12 square before his next square can be planted.

Knowledge of addition and probability, and just pure luck are all part of winning this game.

Figure 9.31

12	4	9
8	6	2
10	3	7

CLIMB UP AND DOWN THE LADDER

And win a point for your team and a star for your forehead (Figure 9.32).

CLIMB UP, SLIDE DOWN

Bulletin board game – boys versus girls. Teacher changes the numeral or the sign on the balloon as different players compete (Figure 9.33).

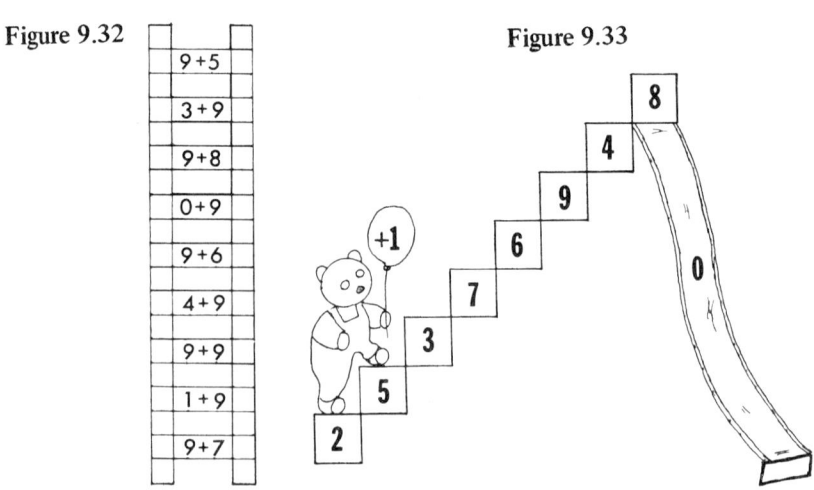

Figure 9.32

Figure 9.33

It's Your Turn – Games

ADD – SUBTRACT – SPELL (Figure 9.34)

ACROSS   DOWN

1. 4 - 1   1. 2 + 0
2. 8 - 2   2. 5 + 2
3. 5 - 4   4. 7 + 1
5. 9 - 4   5. 2 + 2
7. 10 - 1   6. 7 + 3
8. 7 - 7

Figure 9.34

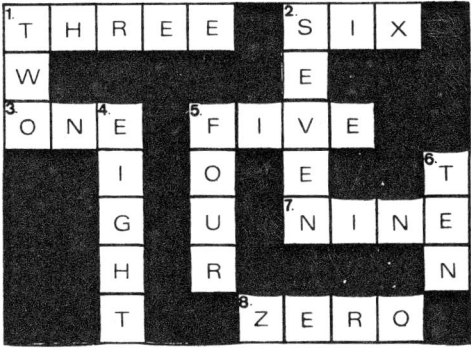

WHIZ-BANG COLUMN ADDITION

Directions for use:

Five addends are selected, each from a different row and column. When these are added the sum will always be 1969, the year men first landed on the moon (Figure 9.35).

Example:

Select	E1	231
	C3	391
	A5	561
	D2	307
	B4	479
		1969

If subtraction is practiced between any two rows, or any two columns, the differences are always the same.

Figure 9.35

Card 1

	1	2	3	4	5
A	180	265	362	464	561
B	195	280	377	479	576
C	209	294	391	493	590
D	222	307	404	506	603
E	231	316	413	515	612

Use Card II and discover the date of another great discovery (Figure 9.36). Directions for constructing similar charts:

1. Select any numbers for Column 1, each succeeding number greater in value.

2. Select any numbers for Row 1, each number greater in value.

3. Find the differences between each of the pairs of consecutive numbers in Column 1. Use this pattern of differences for determining the numbers in the remaining columns. (Notice that a row pattern would have produced the same results.) The column differences for Card I are 15, 14, 13, and 9; for Card II, they are 12, 17, 8, and 14.

4. Use the addends along the diagonals to guide you to the number you have selected for your final sum.

Working out charts of their own birthday year is an engrossing enrichment activity for capable students.

Figure 9.36

Card 2

	1	2	3	4	5
A	12	224	285	392	450
B	24	236	297	404	462
C	41	253	314	421	479
D	49	261	322	429	487
E	63	275	336	443	501

A SQUARE MAN (Figure 9.37)

How many different squares does the Square Man have?

It's Your Turn – Games

Head	5
Neck	1
Arms	4
Legs	4
Body	14
	28

Figure 9.37

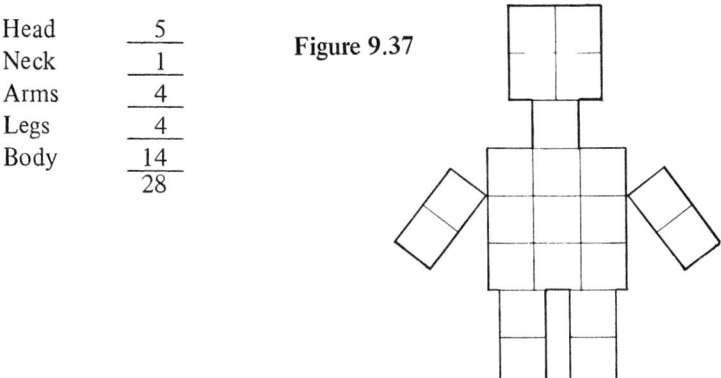

TRUCKS LOADED WITH RECTANGLES (Figure 9.38)

How many different rectangles are in each truck? Hint for counting Truck 3: First count the rectangles that have an area of 12 square units, then the rectangles of 10 square units, and so on down to those of 1 square unit.

Can you find at least;

1. (6)

2. (12)

3. (70)

Figure 9.38

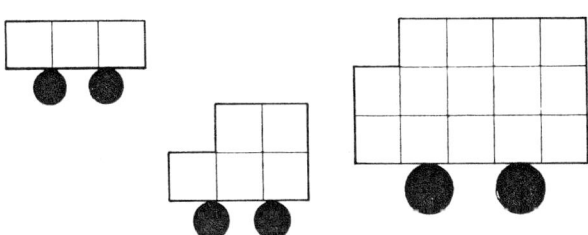

MAKE AN EQUATION – UP – DOWN – ACROSS – DIAGONAL

Move in any direction – horizontally, vertically, or diagonally, upwards or downwards – one square at a time. Form an equation with the numbers you have selected by your moves (Figure 9.39).

Samples:

9 - 4 = 5	35 - 25 = 10
9 - 3 = 6	25 + 10 = 35
9 + 54 = 63	(14 + 8) - 10 = 12
9 x (3 + 6) = 81	[(21 + 56) - (2 + 54 + 6 + 9)] x 2 = 12
27 - (2 - 0) = 25	2 x 7 = 14

Figure 9.39

9	3	6	81	1	38	48	86
54	4	9	3	8	4	5	35
63	58	5	27	2	10	25	33
72	70	20	7	16	0	2	64
58	62	14	8	35	3	60	4
14	8	10	12	6	1	19	26
28	7	4	30	41	13	7	30
21	56	2	54	6	9	2	12

EQUATION MIX-UP

Squares with numerals from 0 to 20 are mixed up in a bag. Each player draws three to begin the game and one at each subsequent turn. The object is to make addition equations. For example, squares showing 3, 2, and 5, may be discarded. The player with the least number of squares left is the winner. The game may be played as a subtraction exercise, or a similar game may be devised to suit multiplication and division equations.

CAN YOU BUILD NUMBERS?

Each child in the class is supplied with a set of digits from 0 to 9. As the teacher calls out the numbers, the child builds them from his set. He may be asked to make 52, or 502, or 520; to show 60, and then 3 more than 60; to build 416, and then 10 more than this number, or 100 more.

GRAND TOTAL – DISCOVER AND EXPLAIN (Figure 9.40)

1. Add rows and columns. What do you discover about the lower right-hand corner? (This is a similar procedure to recording team scores in bowling leagues.)

2. Subtract *sum* of Row 1 and *sum* of Column 1. 39 - 21 = 18. Subtract *sum* of Row 3 and *sum* of Column 3. 63 - 45 = 18. Compare these answers. Explain.

3. Compare *sum* of Row 2 and *sum* of Column 2. Explain.

It's Your Turn – Games   *213*

Figure 9.40

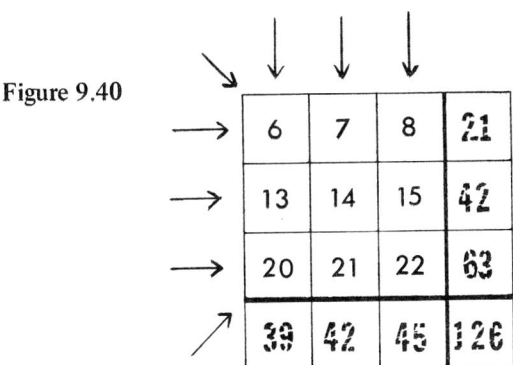

Add the 4 x 4 array in as many directions as possible. Look for patterns and explain. The arrows suggest different addition problems (Figure 9.41).

Figure 9.41

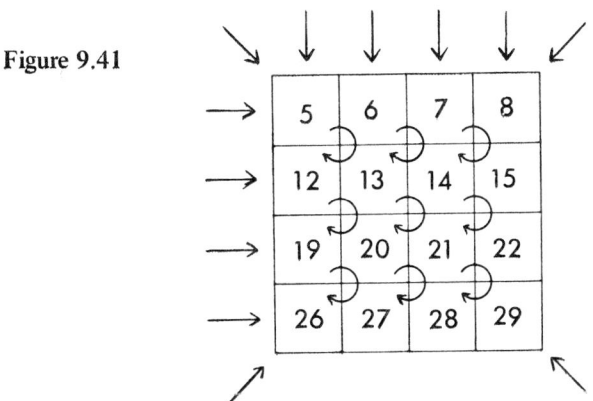

EXTENSIONS OF GRAND TOTAL (Figure 9.42)

Figure 9.42

	+ →		
−	5	4	9
↓	2	3	5
	3	1	4

	× →		
÷	5	4	20
↓	2	3	6
	$2\frac{1}{2}$	$1\frac{1}{3}$	$3\frac{1}{3}$

	÷ →	
↓	64	8
	4	1

+		
$2\frac{1}{2}$		5
4	$4\frac{1}{2}$	

−		
	3.1	2.9
	2.1	
1.5		

×		
0.5		2.0
0.5	0.6	

×		
$\frac{4}{5}$	$\frac{5}{8}$	
$\frac{5}{6}$	$\frac{6}{10}$	

+ Base 5		
4		12
2	101	

+ Mod 5		
4	4	
3	2	

CHECKER MOVE: LEAST COMMON MULTIPLE
(OR LEAST COMMON DENOMINATOR)

Find the least common denominator for 1/4 and 1/6, or the L.C.M. for 4 and 6.

Use the counting chart or number line and checkers to give the atmosphere of a game. The Red checker must always move 4 jumps at a time, and the Black checker must move 6. Each time a checker is moved it must pass the other checker even if this involves several moves at one turn. The object is to guess and write down the number where the two checkers will first land on the same square (Figure 9.43).

Moves:

1. Red moves to 4.   4. Black moves to 12.
2. Black moves to 6.   5. Red moves to 12.
3. Red moves to 8.   The *L.C.M.* of 4 and 6 is *12*.

Discussion should reveal that Red's stops are multiples of 4 (Table of 4 on Multiplication Chart), and that Black's are multiples of 6. Since 4 needed three jumps to reach 12, but 6 needed only two jumps, when finding the least common denominator of two or more fractions, the quickest way is to use the greatest of the denominators and test its multiples by the other denominators involved.

Figure 9.43

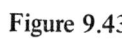
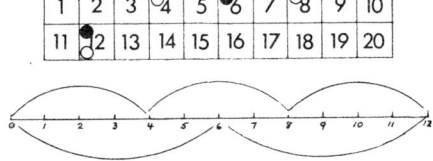

CROSS PUZZLES

Add adjacent arms and write answer between them. Then add across a diagonal and write the answer in the center. Add all the arms of the cross. Where is this answer? Add all the corners. Is this answer predictable? (Figure 9.44)

Figure 9.44

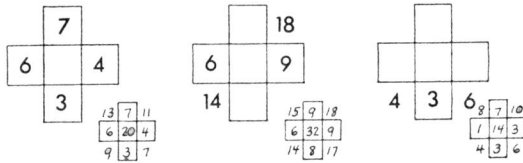

SECRET WORDS (Figure 9.45)

Code:

Figure 9.45

1	2	3	4	5	6	7	8	9	10
r	m	o	c	d	e	g	i	a	n

1. 10 - 3 = **7**   Letter **G**
   11 - 8 = ___   Letter ___
   21 - 18 = ___   Letter ___
   13 - 8 = ___   Letter ___

2. 28 - 26 = ___   Letter ___
   17 - 8 = ___   Letter ___
   13 - 6 = ___   Letter ___
   17 - 9 = ___   Letter ___
   12 - 8 = ___   Letter ___

The secret word is _____   The secret word is _____

Figure 9.46

a	b	c	d	e	f	g	h	i	j	k	l	m	n	o	p	q	r	s	t	u	v	w	x	y	z
1	2	3	4	5	6	7	8	9	10	11	12	13	14	15	16	17	18	19	20	21	22	23	24	25	26

Variation: Use two long rows of squares to include the 26 letters of the alphabet. Make up problems that spell out short sentences, as: "You did a good job today." This visual strip may be used over and over (Figure 9.46). Be sure to include the spelling words for the week when choosing the problems. Also use multiplication and division problems.

Figure 9.47

$\frac{1}{8}$	$\frac{1}{7}$	$\frac{1}{4}$	$\frac{1}{2}$	$\frac{5}{8}$	$\frac{3}{4}$	1	$1\frac{1}{3}$	$1\frac{2}{3}$	2	$2\frac{7}{8}$	3	$3\frac{1}{5}$	$3\frac{2}{5}$	$3\frac{3}{5}$	4
A	B	C	D	E	F	G	H	I	J	K	L	M	N	O	P

FRACTION CODE

Work the problems to find the dog (Figure 9.47).

1. 3-3/8 + 5/8 =
2. 4 - 2/5 =
3. 5-1/10 - 1-1/2 =
4. 6-3/4 - 6-1/4 =
5. 9-7/8 - 6-7/8 =
6. 1 - 3/8 =

Make up problems for Collie, Beagle, Afghan, Basenjii, Cockapoo, and Pekingese.

10.

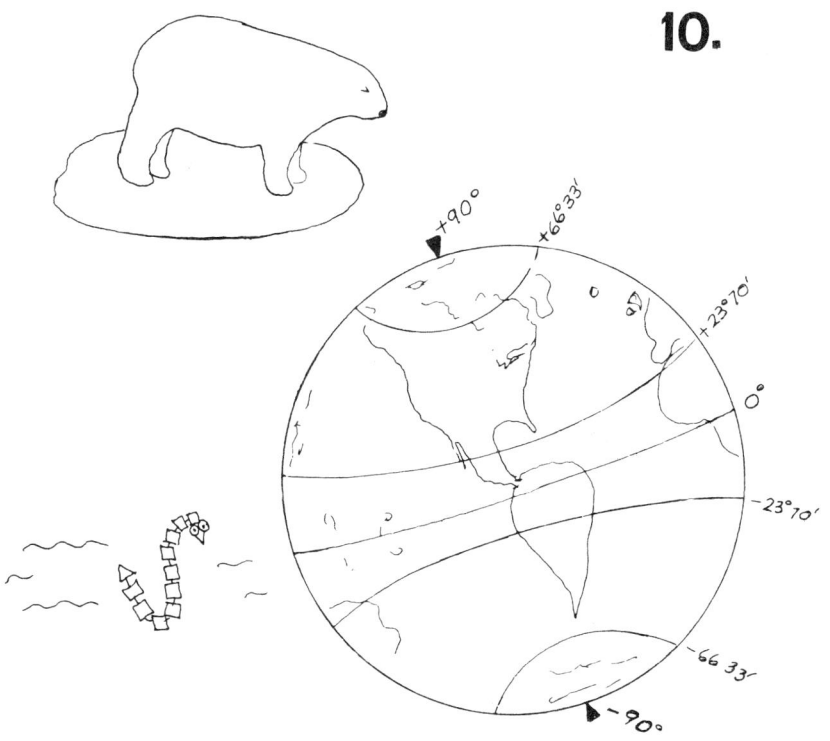

10

 **Odds And Ends**

—locate a school, graph an equation, let a computer think for you, make a point

These extensions are "odd" only in the sense that they do not fit into the other categories, but are sensible because they bring this book to an "end."

OTHER BASES

A few ideas are collected here to show the use of squared paper for exploring and charting other base numerals. The Secret Number Cards, Base 2, are not new, but as an exciting guessing game are worth including, and they do present a practical way of gaining an understanding of the binary base. Intermediate pupils are capable of figuring out which numerals should be placed on each card, and soon perceive the patterns. When the small slotted Computer Cards are used in conjunction with the Secret Number Cards, they will read your friend's mind and show you the number he is thinking of.

For a *surprise*, use a right angle slide and change that ever reliable counting chart into a change-a-base counting chart for every base from 2 to 10.

COORDINATES

Personal involvement in real situations has great value in the learning game. If these situations are indeed games, they will be received and played enthusiastically, even if mathematical ideas are attached as riders.

Tic-Tac-Toe is a simple game, not over exciting until a 4-in-a-row variation is played on a coordinate grid by sixth-graders, boys versus girls. While strategy is being formulated and decisions made, the boys and girls are learning how to locate points by rectangular coordinates. This activity is excellent for introducing ordered pairs of numbers, for finding positions in a plane, and for emphasizing the importance of order.

Another technique, equally effective for starting a lesson on coordinates, is to give each child a number name, an ordered pair, that locates him in the room. A child's positional name, (3, 2), which shows he sits in Row 3, Seat 2, is a 2-dimensional name.

The four dimensions of measurement, length, width, height, and time, helped put men on the moon. These dimensions may be simply explained by making a date to meet a friend. "Meet me at the school on Main Street (length), at the intersection of Grand Avenue (width), in the library on the 3rd Floor (height), at 1:30 P.M. this afternoon (time)."

Only length and width are necessary to locate individual children in the classroom. *Height* need not be considered, since everyone is on the same plane, nor *Time* since all are there together at the same day and hour.

Where Do You Sit?, is a fun activity. Children use their number names to graph their seat positions on a coordinate grid. A bonus is a better understanding of linear equations. For $x + y = 5$, a line appears across the graph, and across the room. Inequalities $x + y < 5$, and $x + y > 5$ appear on particular sides of the graph line and room line.

The grid is essential for graphing all types of equations, including functions and simultaneous equations; for plotting ratios and all ordered pairs of numbers; for demonstrating reflections and symmetry; for map and other scale drawings; and to determine the distance between two points in a plane.

Fractions contain ordered pairs of numbers, denominators and numerators. A set of equal fractional numbers may be produced on a coordinate graph. This use of the coordinate plane has been explained in Chapter 8, Fractions.

PROBABILITY

Mathematics is termed an exact science, yet probability, estimation and logic make up a great part of the mathematics of the business world. Selling, buying, construction contracting, insurance, entertainment, medical, military and other science research are but a few fields where statistical techniques are employed. Before Anaheim was selected as the site for Disney's famous amusement park, the probability for its success was thoroughly checked out by statistical methods.

Investigations of probability, known as samples, may be recorded on the 100-square grid, and these records should help verify estimated chances. Although only a few examples are included here, there will be many opportunities where the 100-square will be useful to chart frequencies, classify data, and by graphing information to study the basic ideas of probability.

When this book was begun, the title selected was, "The 100-square – The Poor Man's Visual Aid." As the suggestions for its use literally added up and began to multiply, the title was changed to "1,001 Ways To Use The 100-Square."

But just as the first few countable raindrops become the downpour and the seemingly endless shower, so the uses of the 100-square are indefinite in quantity, and inexhaustible. Every idea given will probably suggest to the reader another, excellent, practical application that has been omitted here.

The title, "1,001 Uses Of The Hundred Square," is merely descriptive, as are the words "millionaire" or "centipede." 1,001 best describes how versatile the 100-square is when used to provide mathematics learners with visual experiences.

Figure 10.1

CHANGE-A-BASE COUNTING CHART

Color code the right angles or move a right-angle slide over a Base 10 Counting Chart to quickly create counting charts for all bases from 2 to 10. Find a Base 2 Counting Chart inside the zzzz slide; a Counting Chart for Base 5 inside

Odds And Ends 221

the oooo; etc. (Figure 10.2). Where should the slide be moved in order to enclose a counting chart for Base 6? for Base 4?

Figure 10.2

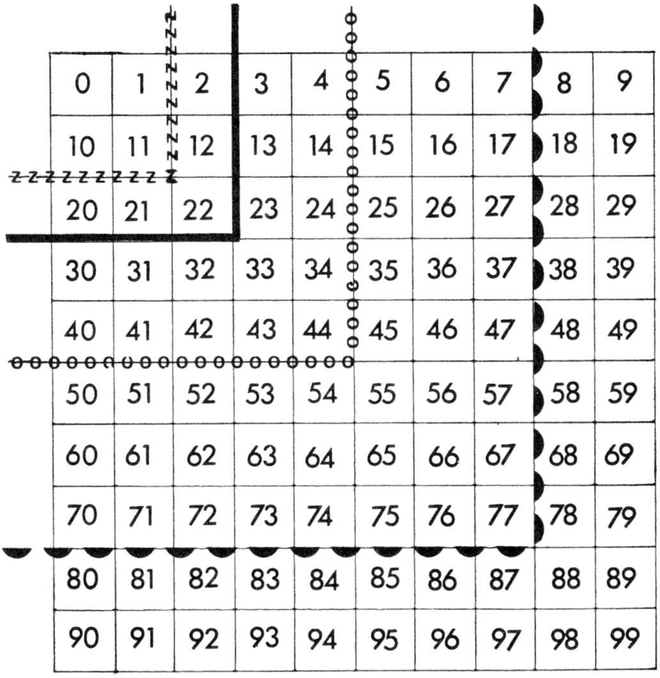

0	1	2	3	4	5	6	7	8	9
10	11	12	13	14	15	16	17	18	19
20	21	22	23	24	25	26	27	28	29
30	31	32	33	34	35	36	37	38	39
40	41	42	43	44	45	46	47	48	49
50	51	52	53	54	55	56	57	58	59
60	61	62	63	64	65	66	67	68	69
70	71	72	73	74	75	76	77	78	79
80	81	82	83	84	85	86	87	88	89
90	91	92	93	94	95	96	97	98	99

BASE 5 COUNTING, ADDITION, AND MULTIPLICATION CHARTS

Children who have had experience in constructing and using counting and operational charts for the decimal system, will be able to apply their organizational knowledge to developing charts for other bases (Figure 10.3). Comparison of these various base charts provides much insight and understanding of place value, expanded notation, inverse operations, laws of operations which might include additive and multiplicative inverses, and ability to change easily to another base and operate within that base. For further enrichment, be sure to encourage the students to explore area, odd and even numbers, primes and composite numbers, and square numbers in bases other than 10. Include work in bases higher than 10, and involve the students in the problem of devising symbols for the extra digits needed. Interested students may construct charts for Base 12 or Base 20.

Figure 10.3

1	2	3	4	10
11	12	13	14	20
21	22	23	24	30
31	32	33	34	40
41	42	43	44	100

+	0	1	2	3	4
0	0	1	2	3	4
1	1	2	3	4	10
2	2	3	4	10	11
3	3	4	10	11	12
4	4	10	11	12	13

×	0	1	2	3	4
0	0	0	0	0	0
1	0	1	2	3	4
2	0	2	4	11	13
3	0	3	11	14	22
4	0	4	13	22	31

BASE 5 PLACE VALUE

Show that 112_5 is equal to 32_{10} by counting the squares in Base 10 (Figure 10.4).

Figure 10.4

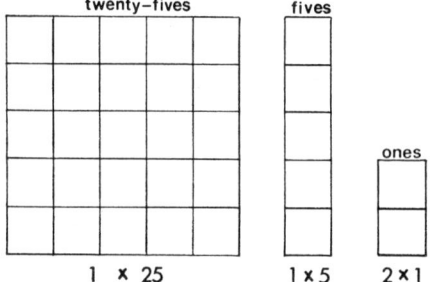

CONVERT BASE 10 TO BASE 5

Use a Base 5 abacus having each rod or spindle marked with the value of the place. Place beads on the rods to equal the value of the Base 10 number. Record the number of beads and the Base 5 number on a chart (Figure 10.5). The abacus below shows $331_{10} = 2311_5$. In expanded notation:

$$(2 \times 125) = 250$$
$$(3 \times 25) = 75$$
$$(1 \times 5) = 5$$
$$\underline{(1 \times 1) = 1}$$
$$2311_5 = 331_{10}$$

Figure 10.5

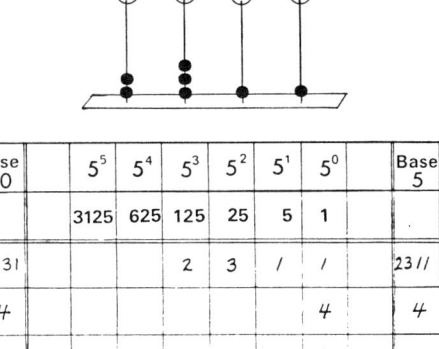

Base 10	5^5	5^4	5^3	5^2	5^1	5^0	Base 5
	3125	625	125	25	5	1	
331				2	3	1	2311
4						4	4
8					1	3	13
81				3	1	1	311
702			1	0	3	0	2	10302
6250		2	0	0	0	0	0	200000
25					1	0	0	100

Experiment with a Base 7 abacus. The place value, represented by the rods from right to left, will be 1, 7, 49, 343, 2401, and 16,807.

SECRET NUMBER CARDS (Figure 10.6)

1. Have someone think of a number from 1 to 63. (In a class situation turn your back while a child writes a number on the chalkboard for all the class to see, and then erases it.)

Figure 10.6

A

1	3	5	7
9	11	13	15
17	19	21	23
25	27	29	31
33	35	37	39
41	43	45	47
49	51	53	55
57	59	61	63

B

2	3	6	7
10	11	14	15
18	19	22	23
26	27	30	31
34	35	38	39
42	43	46	47
50	51	54	55
58	59	62	63

C

4	5	6	7
12	13	14	15
20	21	22	23
28	29	30	31
36	37	38	39
44	45	46	47
52	53	54	55
60	61	62	63

	D		
8	9	10	11
12	13	14	15
24	25	26	27
28	29	30	31
40	41	42	43
44	45	46	47
56	57	58	59
60	61	62	63

	E		
16	17	18	19
20	21	22	23
24	25	26	27
28	29	30	31
48	49	50	51
52	53	54	55
56	57	58	59
60	61	62	63

	F		
32	33	34	35
36	37	38	39
40	41	42	43
44	45	46	47
48	49	50	51
52	53	54	55
56	57	58	59
60	61	62	63

2. Discover what this number is by showing the cards in alphabetical order and asking, "Is your number on this card?" Whenever the answer is "yes" for any card, remember the first number on the card. The sum of these first numbers on the "yes" cards is the secret number.

3. The cards are easily made by putting 1 on Card 1(A), 2 on Card 2(B), 3 on Cards 1 and 2 (since $1 + 2 = 3$), 4 on Card 4(C), 5 on Cards 4 and 1 (since $4 + 1 = 5$), 6 on Cards 4 and 2 ($4 + 2 = 6$), 7 on Cards 4, 2, and 1 ($4 + 2 + 1 = 7$), 8 on Card 8(D), 9 on Cards 8 and 1 ($8 + 1 = 9$), etc. After the Cards have each been well started in this manner ask the pupils if they can discover the pattern for each card. They may describe the pattern sequences and their relationships as: "On Card A (the 1's card), write 1 numeral and then skip 1; on Card B (the 2's card), write 2 numerals and then skip 2; on Card C (the 4's card), write 4 numerals then skip 4." They find that the sequence for the 8's card, the 16's card and the 32's card follow a similar pattern, and the cards may be quickly filled in. Note the role of Base 2 on these Secret Number Cards.

Base 2 Computer Cards

Small computer cards will do all the thinking for you. Using a felt pen, number index cards from 1 to 63. Six holes are punched across the tops of all the cards. Some holes are then cut further to form slots. These slots represent 1 (or yes) in the Base 2 system, while the holes represent 0 (or no). Beginning at the right, the holes and slots represent the Base 2 places of 1, 2, 4, 8, 16, and 32. Since 20 in Base 10 is 10100 in Base 2, slots on this card would be cut only in the 16 and the 4 places ($16 + 4 = 20$). Figure 10.7

Observe the model below which illustrates how the card numbered in Base 10 is cut to show 110011 in Base 2.

$51_{10} = (1 \times 32) + (1 \times 16) + (0 \times 8) + (0 \times 4) + (1 \times 2) + (1 \times 1) = 110011_2$

Figure 10.7

Odds And Ends 225

When the Computer Cards have been properly punched and slotted to represent the Base 2 numbers which correspond to the Base 10 numbers written on them, they are ready for use. These Computer Cards are used in conjunction with the Secret Number Cards (Figure 10.6). Stack them and hold them in your hand. As the Secret Number Cards are shown, stick a knitting needle through the 1's place in the stack of Computer Cards. If the answer to the question, "Is your number on Card A?," is "Yes," remove all the "no" cards from the needle but retain the "yes" cards in your hand. As Card B is shown, stick the needle through the 2's place in the retained cards. If the answer is, "Yes," throw out the "no" cards and retain the "yes" cards in your hand. If the answer for Card C is "No," throw out the "yes" and retain the "no" cards. Continue in like manner for each place in turn until all the Secret Number Cards have been shown. You will be left with one Computer Card in your hand — the secret number.

ELECTRONIC COMPUTERS (Figure 10.8)

Figure 10.8

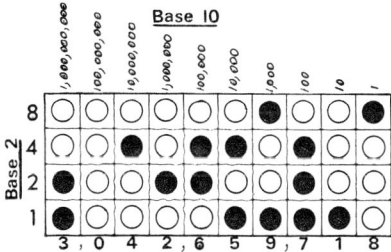

Electronic Computers use a combination of Base 2 and Base 10. The lights in each row represent a Base 2 number. The place value of each column is Base 10.

MULTIPLYING POWERS OF 2 (Figure 10.9)

Use a double number line to show multiplying powers of the same base.

Example 1:

$2^3 \times 2^4$, or 8 x 16. 2 is the base and 3 and 4 are the exponents.

$2^3 = 2 \times 2 \times 2 = 8$ $2^4 = 2 \times 2 \times 2 \times 2 = 16$ $8 \times 16 = 2^3 \times 2^4 = 2^7$

To multiply 8 x 16 on the double number line: (1) put fingers on 8 and 16 on the the bottom line; (2) run fingers up to the top line to find the exponents 3 and 4; (3) add these exponents and get the sum of 7; (4) run across the top line to ⁷ and down to 128, the product of 8 x 16.

Example 2:

Multiply 2048 x 16. Locate the factors, 2048 and 16 on the bottom line. Directly above them find 11 and 4 on the top line. Add, and get 15. Below 15 is the product 32,768. 2048 = 2^{11}, and 16 = 2^4. Their product is 2^{15}.

Figure 10.9

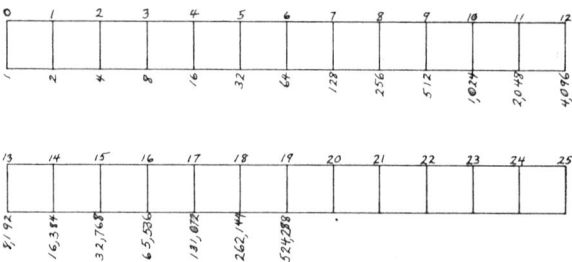

To construct this Base 2 double line chart multiply each product place on the bottom line by 2 to find each succeeding product.

Similar charts may be made for other bases. For Base 5, start with the 1's place, multiply by 5 to obtain the 5's place, use 5 again as a factor to obtain the 25's place, use 5 the third time as a factor to obtain the 125's place, etc. In any base, 1 is the base to the zero power.

COORDINATES: A TRIP TO SCHOOL

The teacher tells a story of her trip to school. At various points she places small pictures of animals, trees, etc., that she saw on the way. Pupils give the ordered pair of numbers (coordinates) that locates the corner where she saw each animal. *Example:* She saw the dog at (1,1), and the turtle at (1,2). Figure 10.10

TIC-TAC-TOE ON THE COORDINATE GRID

The game is played by two teams, girls versus boys. The teacher places red or black checkers as alternate players call out ordered pairs of numbers. The first team to get four checkers in a row, diagonal, vertical, or horizontal, wins the game (Figure 10.11). Different rules may be stated for each game: game may be played in only one particular quadrant, or on the entire coordinate chart; outer boundaries may or may not be used; five in a row may constitute a win; et cetera. This is an excellent game for introducing coordinates.

Odds And Ends   227

Figure 10.10

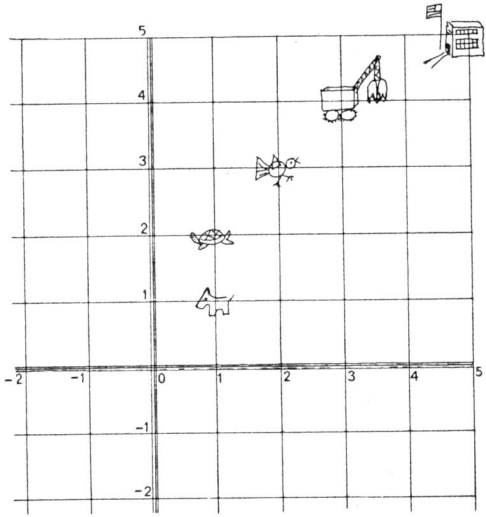

Figure 10.11

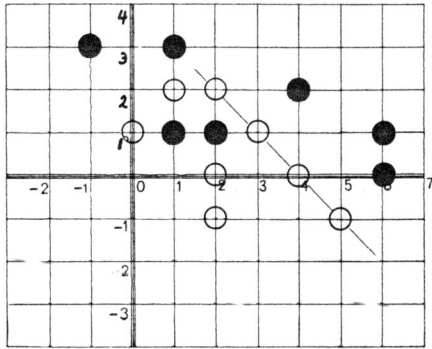

MAP COORDINATES (Figure 10.12)

What town is located at each of the following points?

1) 2A   2) 5A   3) 6F   4) 7-1/2D

Give the points where the following towns are located:

5) Encino   6) Sun City   7) Silver Bell   8) Las Vegas

Note: Fig. 10.12 is a roughly drawn map — not valid.

Figure 10.12

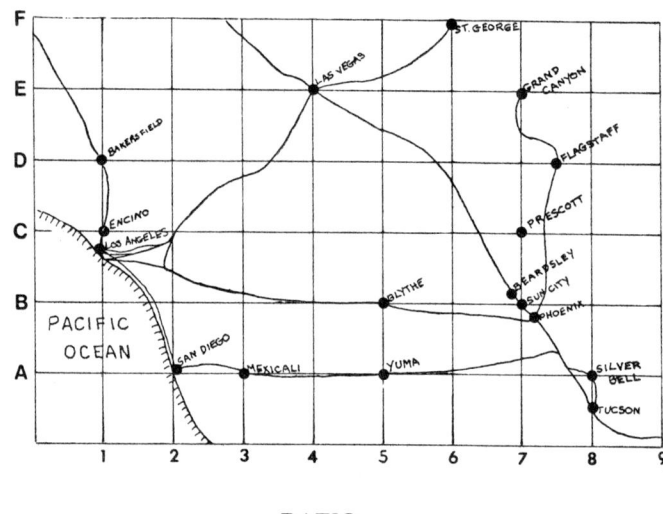

RATIO

Complete the distance ratio chart and use as a key for the map (Fig. 10.13). (Key: 1" = 100 mi.)

Figure 10.13

Map	Scale					
1	1½	2	2½	3	3½	4
100						

SYMMETRY AND REFLECTION ON A COORDINATE GRID (Figure 10.14)

1. Graph these points:

Quadrant I

A	(+1, +4)	H	(+3, +2)
B	(+2, +3-1/2)	I	(+3, +1)
C	(+3, +4)	J	(+2, +1-1/2)
D	(+3, +3)	K	(+1, +1)
E	(+4, +2-1/2)	L	(0, +2-1/2)
F	(+5, +4)	M	(+1, +3)
G	(+5, +1)		

Odds And Ends

2. Complete, then graph these reflective points:

Quadrant II				Quadrant III		
A'	(-1,+4)	H'		A"	(-1,-4)	H"
B'	(-2,+3-1/2)	I'		B"	(-2,-3-1/2)	I"
C'		J'		C"		J"
D'		K'		D"		K"
E'		L'		E"		L"
F'		M'		F"		M"
G'				G"		

3. Draw these segments:

$\overline{AE}$	$\overline{EG}$	$\overline{A'E'}$	$\overline{A"E"}$
$\overline{BC}$	$\overline{KE}$	$\overline{B'C'}$	$\overline{B"C"}$
$\overline{CD}$	$\overline{HI}$	$\overline{C'D'}$	$\overline{C"D"}$
$\overline{EF}$	$\overline{JI}$	etc.	etc.
$\overline{FG}$	$\overline{LK}$		

4. Name coordinates and segments needed to graph the reflection in Quadrant IV.

Figure 10.14

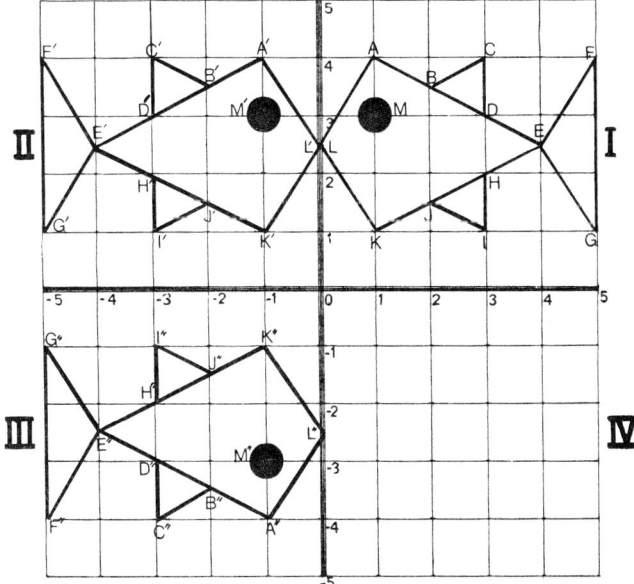

THE GRAB-BAG GAME AND SOLUTION SETS

Two bags are labeled ☐ and △. Two pupils draw cards, one a triangular-shaped card, such as △, and the other a square-shaped card, such as ☐. Only the teacher sees their cards, and after determining the sum, writes the equation on the chalkboard; △ + ☐ = 6. The class tries to guess what numeral each child is holding, and these answers are recorded in a T-table. When all possible correct solutions have been given, the grabbers place their cards in the frames so it can be seen who guessed exactly what they were holding. Winner is a grabber next time (Figure 10.15). After several games using frames for the variables, turn the bags to show the labels x and y. Then the equation might be x + y = 5, and guesses are recorded in a table labeled x and y. Primary children learn facts and use the commutative law of addition as they enjoy this game.

Figure 10.15

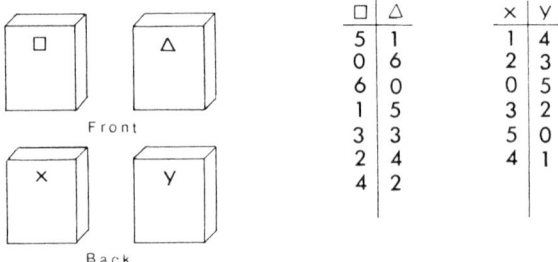

GRAPHING THE GRAB-BAG EQUATION

The equation, x + y = 6, is a linear equation and is represented on the coordinate grid. The solutions, (0,6), (1,5), (2,4), (3,3), (4,2), (5,1), and (6,0), are located and marked on the grid. A connecting line may be drawn (Figure 10.16). Compare this diagonal line with the line made by the 6-sums on an additional chart.

Figure 10.16

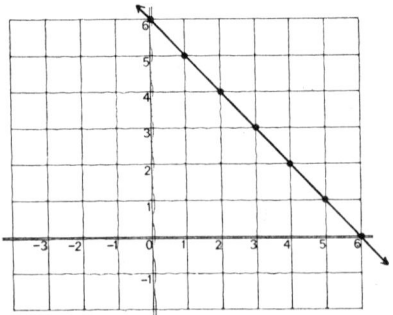

Odds And Ends

WHERE DO YOU SIT?

It is challenging and fun for children to graph the points that represent their positions in the classroom. Arrange their seats in a rectangular array and give them coordinate names. For example, the child who is seated in Row 1, Seat 4, is named (1,4). After all the children have been similarly named, whenever x is used they must think of their first number names, and for y, their last number names. Give directions to add $x + y$, and if the sum is 5, to stand. The standing children will form a diagonal line across the room. They graph their points on a large 100-square grid, and the linear equation, $x + y = 5$, appears on the coordinate plane. Next, the children whose names satisfy the inequality, $x + y < 5$, stand and graph their points, and last, those whose names satisfy $x + y > 5$ graph their positions. (Figure 10.17) Discussion reveals that the line separates the plane into three sets of points, those on the line itself, the points on one side of the line, and the points on the opposite side, and though that only a few points are shown, the plane is an infinite set of points. Algebraically, for any two numbers, one and only one of the three order conditions is true: the numbers are equal, the second number is less than the first, or the second number is greater than the first.

Figure 10.17

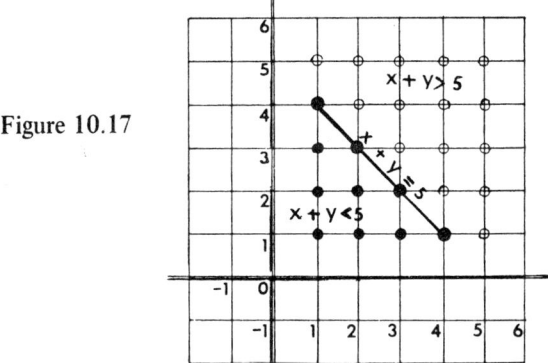

GRAPHING SIMULTANEOUS EQUATIONS

Two equations, $x + y = 5$, and $y = 3$, are acted out simultaneously. Pupils, whose names satisfy the first equation remain standing in their line, while at the same time, pupils whose y, or last name, is 3 stand. This particular second equation forms a horizontal line across the room. One pupil, (2,3), stands in both the diagonal and horizontal lines, and represents the solution that satisfies both equations. This point of intersection shows on the graph. (Figure 10.18) Extended lines on the graph indicate that there may be continuous as well as discrete solutions to the equations.

Figure 10.18

x	y
1	4
3	2
0	5
4	1
2	3
5	0

$x + y = 5$

x	y
0	3
1	3
2	3
3	3
4	3
5	3

$y = 3$

Figure 10.19

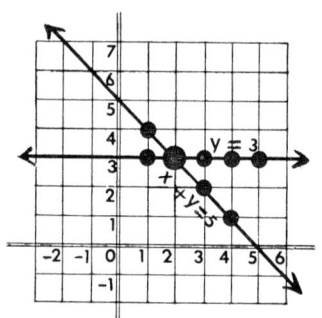

INTEGER MULTIPLICATION PATTERNS (Figure 10.20)

Figure 10.20

A.

		-4	-3	-2	-1	0	+1	+2	+3	+4
X	+4	-16	-12	-8	-4	0	+4	+8	+12	+16

B.

X	-4
+4	-16
+3	-12
+2	-8
+1	-4
0	0
-1	+4
-2	+8
-3	+12
-4	+16

D.

X	+	−
+	+	−
−	−	+

÷	+	−
+	+	−
−	−	+

Figure 10.21

C.-1

X	-4	-3	-2	-1	0	+1	+2	+3	+4
+4	-16	-12	-8	-4	0	+4	+8	+12	+16
+3	-12								+12
+2	-8								+8
+1	-4								+4
0	0								0
-1	+4								-4
-2	+8								-8
-3	+12								-12
-4	+16	+12	+8	+4	0	-4	-8	-12	-16

C-2.

X	-4	-3	-2	-1	0	+1	+2	+3	+4
+4	-16	-12	-8	-4	0	+4	+8	+12	+16
+3	-12	-9	-6	-3	0	+3	+6	+9	+12
+2	-8	-6	-4	-2	0	+2	+4	+6	+8
+1	-4	-3	-2	-1	0	+1	+2	+3	+4
0	0	0	0	0	0	0	0	0	0
-1	+4	+3	+2	+1	0	-1	-2	-3	-4
-2	+8	+6	+4	+2	0	-2	-4	-6	-8
-3	+12	+9	+6	+3	0	-3	-6	-9	-12
-4	+16	+12	+8	+4	0	-4	-8	-12	-16

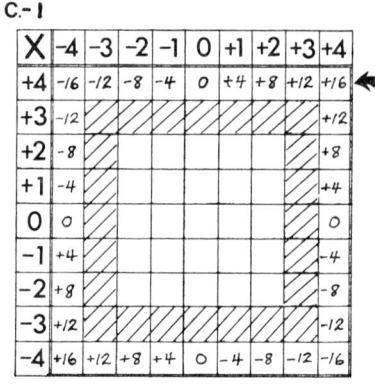

Odds And Ends

A. List integers in a horizontal counting row. Starting at the arrow, multiply each integer by +4, and write the answer in the square under it. Notice the pattern of the signs, and that zero is the focal point for change.

B. Follow a similar pattern for multiplying down a vertical row. Here, too, zero is the focal point of sign change.

C. List the first factors down the left side and the second factors across the top of the chart. Start at the arrow, follow the pattern progression, and fill in the products completely around the square, by multiplying: (1) positive times positive numbers; (2) positive times negative numbers; (3) negative times negative numbers; and (4) negative times positive numbers. Compare with A and B. Next, fill in the shaded row, and continue filling rows around the square until the center zero is in.

D. Form conclusions regarding the multiplication of signed numbers. P x P = P; P x N = N; N x N = P; and N x P = N. (Like signs yield positive answers and unlike signs yield negative answers.)

PROBABILITY — FLIPPING A PENNY TO START A GAME

To start a football game a coin is tossed in order to decide which team will kick-off. Our school team called, "Heads." What is the probability that they will get the choice to kick or receive?

Try this experiment: Flip a penny. Draw up a line unit for heads, or across a line unit for tails. Continue to flip the penny and draw until the graph goes off the chart. Repeat the experiment at least five times. The "ups" (vertical line units) for all five experiments are added, and the "acrosses" (horizontal line units) are likewise totaled, to see if they are equal in number. Normally, the chances of getting heads or tails are equal to 1/2. Trying a long series of penny flips should verify this probability (Figure 10.22). *Variation:* Two different colored wooden cubes (or marbles) are placed in a bag (or other hidden situation). Each child in the class has a chance to grab out a cube. If red, he draws an up unit, and if green, an across unit. After each grab the cube is replaced in the bag. When a line goes off the chart, begin another, and continue until all children have had a turn.

MATH LUCK

Janet is playing a game with her friends. She needs a 2 to get her last counter home and win the game. She tosses two cubes each showing one to six dots on their faces. What is the probability that one of the tossed cubes will show what she needs — a 2-dot face? Make a chart showing all the possibilities? (Figure 10.23) *Suggestion:* Use two wooden cubes, one red and one green, and make dots with a felt marking pen. An oatmeal box, with a hole cut in a taped-on top, is used to toss the cubes easily and quietly. Also, large quiet cubes may be cut from foam rubber.

"A" shows the faces on the red cube and "B" represents the green cube.

Figure 10.22

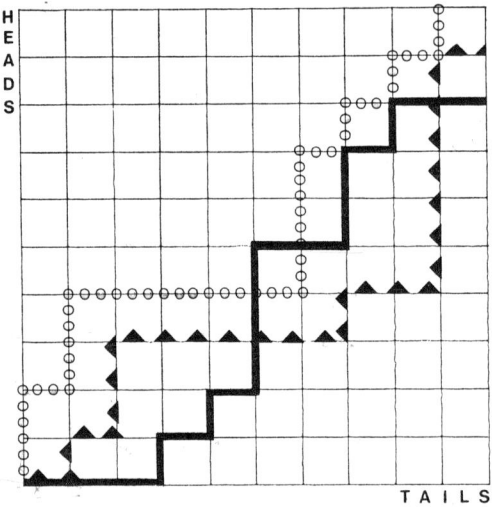

Figure 10.23

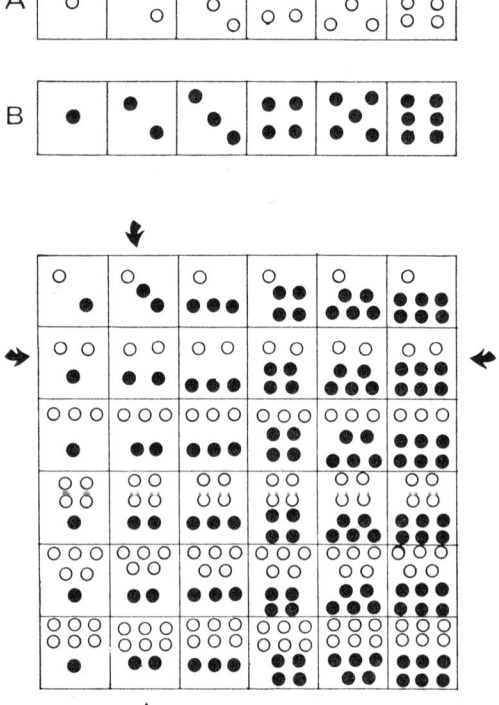

Odds And Ends 235

Numbering the cube tossing distribution chart makes it easier to construct and read. The first number in each ordered pair represents the dots on the red cube and the second number represents the green cube. The chart shows that at least one 2 should turn up 11 times out of 36 tries (Figure 10.24).

Figure 10.24

	CUBE 2					
	1	2	3	4	5	6
1	(1,1)	(1,2)	(1,3)	(1,4)	(1,5)	(1,6)
2	(2,1)	(2,2)	(2,3)	(2,4)	(2,5)	(2,6)
CUBE 1 3	(3,1)	(3,2)	(3,3)	(3,4)	(3,5)	(3,6)
4	(4,1)	(4,2)	(4,3)	(4,4)	(4,5)	(4,6)
5	(5,1)	(5,2)	(5,3)	(5,4)	(5,5)	(5,6)
6	(6,1)	(6,2)	(6,3)	(6,4)	(6,5)	(6,6)

Sample testing: Two cubes, faces numbered from 1 to 6, are tossed simultaneously 36 times. A record is kept of the actual results to compare with the probability chart and the probability number of 11/36.

The H AND T GAME

To play this game: Place two large red wooden beads and two green wooden beads of the same size in a bag. The red beads are worth 10 points each, and the green ones are worth 100 each. Draw out two beads at a time. To win you must get the highest score, and so you hope to get the two 100-beads (or H beads) each time you draw. Each player draws 20 times before the scores are totaled. Each makes a graph of the beads he actually drew each time, and compares his graph with the probability chart. (Beads are replaced after each draw.) Figure 10.25

Now mark a dot on one of the red beads and on one of the green beads. A dot on a bead doubles its value. What are your chances of getting one of these? Try 20 times and make a graph. Compare with the probability chart (Figure 10.26).

Figure 10.25

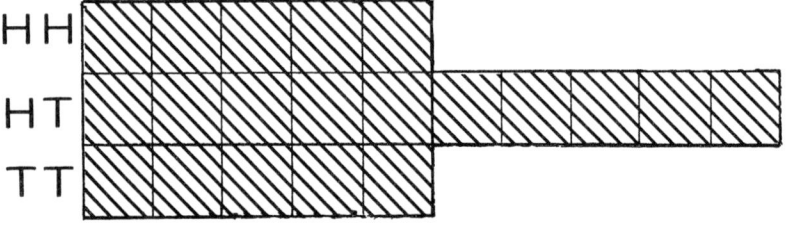

Figure 10.26

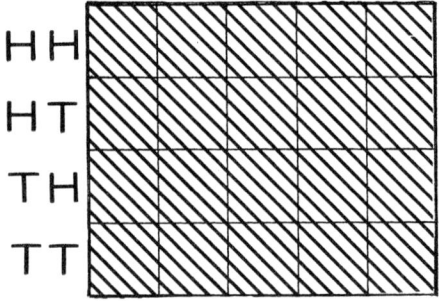

Index

A

Abacus, 90-91, 223
Absolute Value, 62
Addition, 73-78, 88-107, 194-215, 222, 230
    arrow, 76
    chart, 96-97
    column, 76-78, 209-210
    counting chart, on, 76-78
    evens and odds, 78-80
    properties, 101-103
    related facts, 75, 93-94, 105
    tens and nines, 74-76, 80
Algebra, 130-132, 153
Algorithm
    Euclid's, 187-188
    multiplication, 127-130
Array
    bigger squares from small squares, 149
    difficult facts, 119, 130, 138, 149
    distributive law, 126-132, 185
    King's men, commutative law, 123
    primes and composites, 136-138, 149
    rectangular, square, 136-137, 149
Area, 136-137, 142, 146-161, 169
    big boys, 152
    building a pen for pet, 150-152
    cards, 153
    circular region, 159
    cube, around a, 156
    design, 24-25
    fractional regions, 183
    lost, 156-157
    prism, surface, 160-161
    rectangular region, 154-155
    sweeping, 153-155
    triangular region, 154-155
Arrow moves, 58-59, 76
Associative law, 102-103, 134-135
Averages, 107-108
    baseball, 58
    bell curve, 39-40
    equivalent sets, of, 121
    introducing, 120-121
    mean, median, mode, 39

B

Balancing
    arrangement, 23
    equations, 94-95
Bases, other, 61, 214, 218, 220-226
Batter-graph, 44-45
Bell curve, 39
Betweeness, 71-72
Big boy, 152
Bigger squares from small squares, 149
Birthday graph, 34, 36-37
Building a ten, 57-58
Building numbers, 212
Bulletin board ideas, 15, 22, 33, 35, 36, 42
    43, 45, 49, 56-59, 64, 88, 94, 109,
    115-118, 135, 141, 167, 170, 192-193

C

Calendar
    addition, multiplication, on, 103-104
Candles, multiplication facts, 118
Can you build numbers?, 212
Cartesian product, 115
Centigrade thermometer, 165-166
Centimeter, square, 148
Checker move, 214-215
Circular region, area, 159
Commutative law, 67, 73, 101-102, 111, 123
Comparison
    area, perimeter, 150-152
    graphs, 43-45
    puzzle, 186
    subtraction, 56
Composite numbers, 137-138, 149
Computer cards, 223-225
Coordinates, 175-176, 200, 218-219, 226-233
Cotton production, 41-44
Crazy fractions, 188
Criss-cross, 104, 175
Cross products, 115
Cube, 156
Curves, 28-30, 39-40

D

Decimal, 50-51
    fractions, 45-46, 182-183
    number line, 183
Denominator, 174-175, 214-215
Design, 15-31
Diamonds and sparkle, 25
Difference, 21, 35, 210
Dimensions, 142-143, 149, 219
Distance, 62
Distribution, 39, 235
Distributive law, 126-134, 185-186

237

Division, 82, 119, 187-188, 203-206, 213
    chart, 118-120
    exploring, 120-122, 133
    number line, 182
    2-place, 134
    zero, never, 122-123
Divisor, 187
Does the hat fit?, 145
Do it yourself, 172
Dominoes, 92

E

Equations
    linear, simultaneous, 222, 231-232
Equivalence
    class, 169
    fractional, 169-175
    sets, 51-58, 112-113
"Er" suffix, 129
Eratosthenes' sieve, 85-86
Estimation, 54, 62, 97, 128
Euclid's algorithm, 187-188
Even numbers, 78-80
Exponents, 225-226

F

Factoring, 137
Factors, 85, 138, 207
Fahrenheit thermometer, 165-166
Fewer, 52
Fingers
    distributive law, 128
    fifth table, 125
    ninth table, 124
Flashcards, 54, 92-93, 99, 105
Follow the number, 30-31
Formula
    area of
        circle, 159
        rectangle, 154
        triangle, 155
    changing thermometer scales, 165-166
    odd number, 27
    Pythagorean theorem, 164
Four square units, 153
Fractions
    addition, subtraction, 170-177
    decimal, 61, 168-185, 189
    design, 25-26
    distributive law, 185-186
    equivalent, 169, 173-175, 188
    graph, 169
    multiplication chart, on, 173-175, 184, 188
    number line, 61
    slide rule, 63

G

Gallon, 146
Games, 191-215

Games *continued*
    Add-0, 198
    Add-0-Win, 198
    Add, subtract, spell, 209
    Around the block, 197
    Baseball, 199
    Battleship, 200
    Bear went over the mountain, 195
    Big boys, 152
    Can you build numbers?, 212
    Checker move, 214-215
    Climb the ladder, 208
    Climb up, slide down, 208
    Comparison puzzle, 186
    Crazy fractions, 188
    Cross puzzles, 215
    Dominoes, 92
    Equation mix-up, 212
    Extra numeral, 81
    Fact-0, 198
    Facts race, 126
    Ferris wheel, 204
    Follow me, 213
    Follow the numbers, 30-31
    Football, 199
    Fraction-action, 198
    Fraction code, 215-216
    Frame game, 205
    Go, 195
    Grab bag, 230
    Grand total, 212-213
    H and T, 235-236
    Hopscotch, 203-204
    Jumper, the cricket, 191
    Lost area puzzle, 156
    Lost numeral, 86-87
    Lucky tens, 196
    Magic squares, 205-206
    Make an equation, 211-212
    Math words, 202
    Mouse ran up the clock, 203-204
    Multiplying exponents, 225-226
    Multo-0, 198
    Park the cars, 201
    Pebble toss, 206-207
    Picture check, 99
    Plant the corn, 208
    Poison square, 193-194
    Productivity, 198
    Race track, 204-205
    Roll overs, 199-200
    Rule pictures, 196
    Secret number, 223-224
    Secret words, 215-216
    Sign game, 202
    Space trip, 197
    Spill the beans, 207
    Square man, 210-211
    Square nims, 200
    Tic-tac-toe, 206
        on coordinate grid, 226-227

Index

Games *continued*
    Trucks loaded with rectangles, 211
    Whiz-bang column addition, 209-210
    Wolf and little pig, 194-195
    Zip-0's, 197
Graphs, 33-47
    fractions, 169
    introducing a lesson, 41
    linear, simultaneous equations, 231, 232
    where do you sit?, 231
Greater than, 34, 52, 72, 97
Greatest common divisor, factor, 187-188

H

H and T game, 235
Height, 145, 152, 219
Hexagon, 162
Hi-boy, lo-boy, 152
Hypoteneuse, 164-165, 176

I

Identity factor, 111, 124
Inchworm, 141, 147-149
Integers, 62-63, 202-203, 232-233
Inverse operations, 91, 94
Item analysis, 40

J

Jim's Rule, 107-108

K

King's Men, commutative array, 123

L

Lateral faces, 160-161
Lattice multiplication, 138-140
Laws, operations, 17, 67, 101-103, 111, 124-135, 185-186
Length, 154-155, 219
Less than, 34, 52, 72, 97-98
Limits, upper, lower, 129
Line
    design, 28-30
    graphs, 44-46
    segment, 28, 228-229
Linear equations, 230-232
L-o-o-o-ong problem, 131
Lost area, 150-152

M

Magic square, 205-206
Magic sums, 76-78
Map coordinates, 227-228
Math luck, 233
Maxi-midi-mini, 127-129

Mean, median, mode, 39
Measurement, 142-146
Mile, feet in, 146
Minus, 91
Missing links, 19
Money, 177-179
More, 52
Motif, 24-25, 29
Multiplication, 78-85, 109-140, 188, 192, 198, 200-207, 213-215
    arrays, 114, 149
    chart, 118-120, 123, 126, 173-174, 198
    cross-product, 115
    distributive law of, 126-132
    equivalent set, 112-113
    exponents, 225-226
    facts, 78-85, 110-111, 115-119, 123-126
    fractions, 173-175
    integers, 232
    introducing, 112-115
    lattice, 138-139
    number line, 114
    repeated addition, 113
Multiples, 82-86, 197, 214-215
Multipliers, 129-130

N

Napier's bones, 139-140
Negative numbers, 202, 232-233
Noah's Ark, 116
Number, 40-63
    cards, 54
    even, 78-80
    graphs, 37-39
    lines, 60-63, 91, 114-124, 182-183, 197-198
    names, 55, 209
    odd, 27, 37, 79-80, 136-137
    order, 52-54, 67-69, 74
    positive and negative, 202, 232-233
    prime, 83-87, 137-138
    rounding, 62, 74
    sequence, 18-20, 27
    theory, 27
    two-, three-digit, 222
Numerals, 50, 53-56, 69-71, 91
Numerator, 174-175

O

Octagon, 162
Odd numbers, 27, 37, 79-80, 84, 136-137
One off the square, 136
One-to-one correspondence, 28-29, 51-52
Ordered pair, 36, 45-46, 228-229

P

Parabola, 28
Parquetry, 22

Partial products, 128-130
Pave a road, 182-183
Pens for pets, area, 150-152
Percent, 25-26, 38-39, 45-46, 125, 168, 179-182
Penny flip, 233
Pentominoes, 163
Perimeter, 147-152
Pythagorean theorem, 164
Picture flashcards, 99
Place value, 57-58, 71, 104-106, 133, 212 222-223
Plus, 91
Polygons, 161-162
Polyominoes, 163
Positive numbers, 202, 232-233
Powers of two, 225-226
Primes, 82-87, 137-138, 149
Prism, 160-161
Probability, 208, 219-220, 233-236
Progressions, 27
Protractor, 159-160

Q

Quart, 146
Quickie, 103

R

Race track, 126
Ratio, 228, 357
Reciprocal, 184-185
Rectangle, 163-164, 211
Reflection, 228-229
Regrouping, 104-107, 176-177
Roman numerals, 62
Round man may be a square, 131
Rounding numbers, 35, 62
Ruler story, 144

S

School, area, 152
Scale drawing, 157-158
Secret number cards, 223-224
Segments, 28
Sequence, 18-20
Shapes, 17-21
Show-me-strips, 95
Slide rules, 62-63
Slope, 175-176
Snail on a scale, 189
Solution set, 231
Specific sums, 99

Square
  centimeter, 148
  inch, foot, yard, 144-150
  numbers, 27, 136
  root, 85
  root of 2, 165
Squaring
  binomial, 131-132
  numbers ending in five, 136
  tens, 136
Stars and zigzags, 100
Story problems, 41, 62, 95-96, 121-122, 128 134, 161, 182
Subtraction, 73-78, 93-108, 188, 203, 209, 213, 214, 230
  regrouping, 106-107
  fractions, 176-177
Surface area, 160-161
Sweeping, area, 153-154
Symmetry, 18, 23, 29-30, 228-229

T

Tens
  building a, 57-58
  chart, 100
Tessellations, 161-164
Thermometers, 165-166
Thousand, size of, 58
Tic-tac-toe, 206, 226-227, 241
Touch-me-numerals, 56
Triangle, 17, 101, 154-155, 164
Trip to school, 226

U

Units
  four square, 153
  puzzle, 186
  standard, 148-150

W

Whale on a scale, 189
Where do you sit?, 231
Width, 146, 154-155, 219

Z

Zero, 113, 122-123, 129-130
Zigs and zags, 101